P9-CKY-204

DERICO 38-296

MATTER
AND
ENERGY

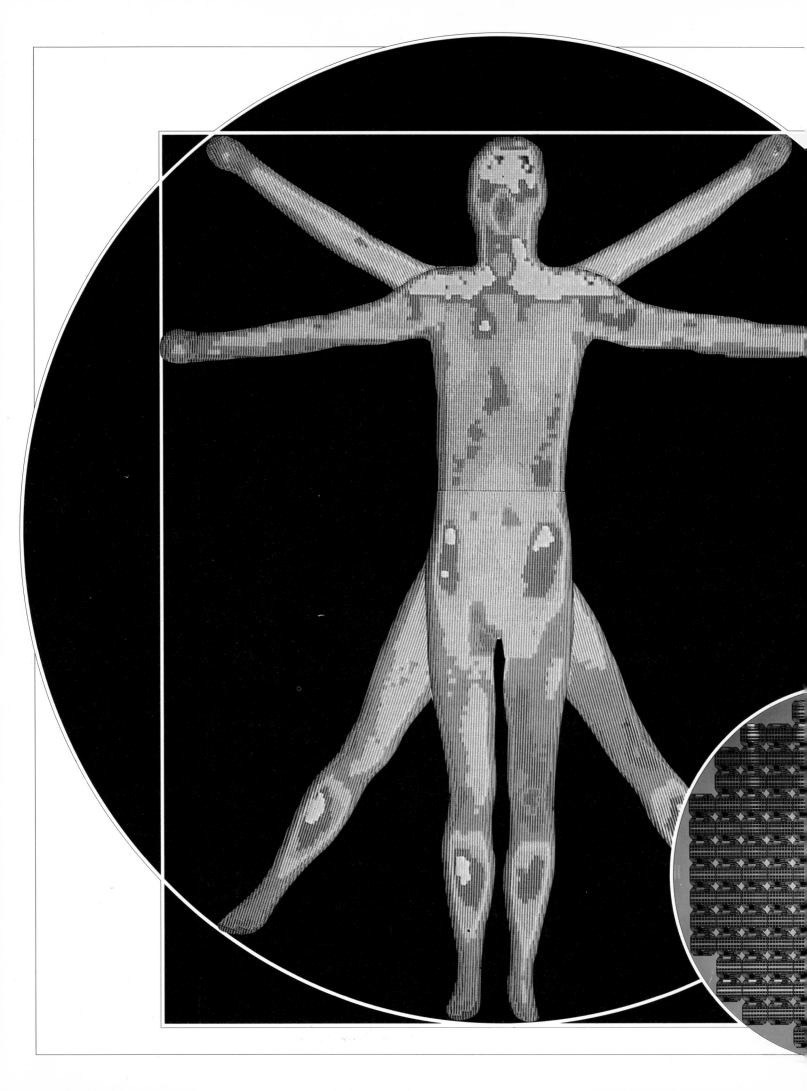

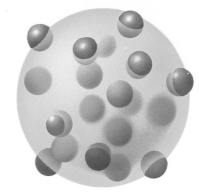

MATTER
AND
ENERGY

Physics in Action

J.O.E. CLARK

OXFORD UNIVERSITY PRESS

New York 1994

SEP '96

Riverside Community College
Library
4800 Magnolia Avenue
Riverside, California 92506

QC 21.2 .C595 1994

Clark, John Owen Edward.

Matter and energy

...TS

Project editor	Peter Furtado
Senior editor	John Clark
Editor	Lauren Bourque
Editorial assistant	Marian Dreier
Art editor	Ayala Kingsley
Visualization and artwork	Ted McCausland/ Siena Artworks
Senior designer	Martin Anderson
Designer	Roger Hutchins
Picture manager	Jo Rapley
Picture research	Jan Croot
Production	Clive Sparling

Planned and produced by
Andromeda Oxford Ltd
9-15 The Vineyard
Abingdon
Oxfordshire OX14 3PX

© copyright Andromeda Oxford Ltd 1994

Text pages 16-47
© copyright Helicon Ltd,
adapted by Andromeda Oxford Ltd

Published in the United States of America by
Oxford University Press, Inc.,
200 Madison Avenue
New York, NY 10016

Oxford is a registered trademark of Oxford University Press

All rights reserved. No part of this publication may be reproduced, stored in a
retrieval system, or transmitted, in any form or by any means, electronic, mechanical,
photocopying, recording or otherwise, without the permission of the publisher.

Library of Congress Cataloging-in-Publication Data

Clark, John Owen Edward
    Matter and energy : physics in everyday life / by John Clark
            160 p.cm. -- (The New encyclopedia of science)
    Includes bibliographical reference (p. 151) and index
    ISBN 0-19-521085-9: $35.00
    1. Physics.   2. Matter   3. Force and energy   I. Title   II. Series.
QL21.2C595   1994
539--dc20   94-16054 CIP

Printing (last digit):9 8   7 6 5 4 3 2 1

Printed in Spain by Graficromo SA, Cordoba

INTRODUCTION

WE LIVE IN A MATERIAL WORLD. Everything is made of matter, and this matter changes when it is acted on by various forms of energy: it moves, it heats up, it expands, it glows, it interacts with other matter. The scientific discipline known as classical physics – the main topic of this book – is the study of matter: what it is, what its properties are, and how it relates to the forms of energy encountered in nature – heat, light, electricity, magnetism and sound, among other forms.

At the start of the 20th century, Albert Einstein and Max Planck set in train an even more fundamental discovery: that on a very small or a very large scale, matter and energy are themselves equivalent and interchangeable. These insights gave rise to a new era in physics, and required the development of the rather different and sometimes bizarre notions of quantum physics, which describes matter and energy on a very small scale.

While physicists have studied all these questions in theory, engineers and technologists have been quick to adopt them in all kinds of practical applications, from transport to microscopes, from information technology to the supply of energy resources. We are surrounded by machines, many of which we take for granted, which rely on quantum phenomena. This book deals then with the basic principles of physics and with their important modern applications.

The great achievement of the classical physicists from the 17th century was to discover the fundamental structure of matter. They came to realize that everything, whether it be a gas, a liquid or a solid, is made of atoms, and that the apparent differences between materials can be explained in terms of the arrangement and behavior of their atoms in the conditions of temperature and pressure normally found on Earth. The thematic

section at the heart of this book therefore begins with an introduction to the states of matter, together with some applications of the principles familiar in modern life. The second key theme – energy and the forces through which it acts – is introduced next, showing how matter can be acted upon in simple ways in order to make machines.

Electricity and magnetism are fundamental, closely related forms of energy, originating in the structure of the atom itself. They allow the transmission of energy conveniently, following principles that were mainly discovered during the 19th century. The applications of this knowledge have fundamentally changed the world in the 20th century, and, through the continuing revolutions in electronics and miniaturization, will continue to do so well into the 21st.

By the end of the 19th century electromagnetism had been shown to be identical in all but energy level to light, heat and other forms of radiation. This shortly led on to the discovery of radio waves and X rays, both with important and familiar applications.

Such electromagnetic radiation, the basic carrier of so much of the energy of our Universe, is a property of the atom itself. The final part of the book, therefore, is devoted to the structure of the atom, with its tiny, dense nucleus orbited by clouds of electrons contained in "energy shells" around them. As scientists delved into these fundamental particles and their arrangement in stable atomic forms, they demonstrated the truth of Einstein's thesis that matter and energy are one, and developed completely new ways of describing matter, interaction, space and force. Even as they did so, the introduction of new technologies – from the nuclear power station to the medical body-scanner – showed the effectiveness of their insights.

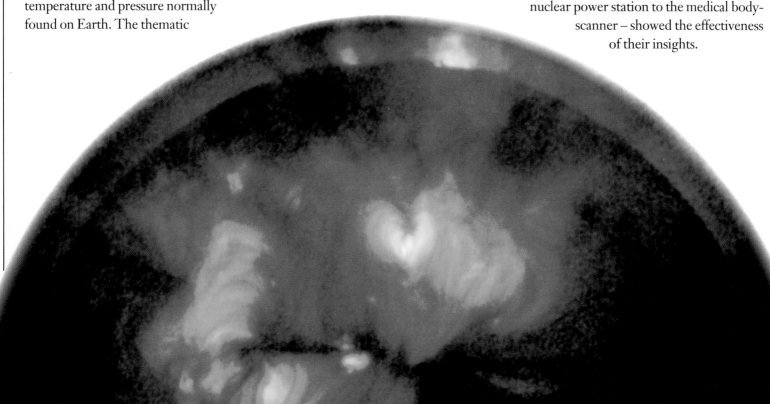

THIS BOOK aims to make all this information available to the whole family, from students studying for exams and projects to adults wanting to bring their scientific knowledge up to date. To achieve this, the book is organized in such a way as to provide readers with a quick answer to a specific query, or allow them to follow a more detailed account of a particular topic.

At the heart of the book is a 96-page thematic section, made up of 48 major narrative topics, each one richly illustrated to tell the story of a central theme of the book. The strong graphic presentation and the style of writing are designed to make this section the ideal point of departure for the less well-informed reader. Sets of Keywords highlighted on each topic spread point the reader to the second major section, a 32-page alphabetic mini-encyclopedia of the subject, which contains some 400 entries. This section, too, leads the reader back to the thematic topics.

No part of modern science can be separated cleanly from other fields. Physics merges into astronomy on the one hand, chemistry on the other and earth sciences in other respects. It has also illuminated aspects of modern biology, and without physics, modern computing and information technology would be impossible. The Knowledge Map, immediately following this Introduction, maps out the entire field of modern science, shows how each area of science interacts with another, and defines the major fields. This is followed by a Timechart, which traces the development of the subject through great discoveries.

Finally, to ensure that the volume is of genuine value for reference as well as for recreational browsing, the Factfile provides a wealth of hard data, tables and statistics, which cover a range of topics introduced earlier.

KNOWLEDGE MAP
Key Fields of Modern Science

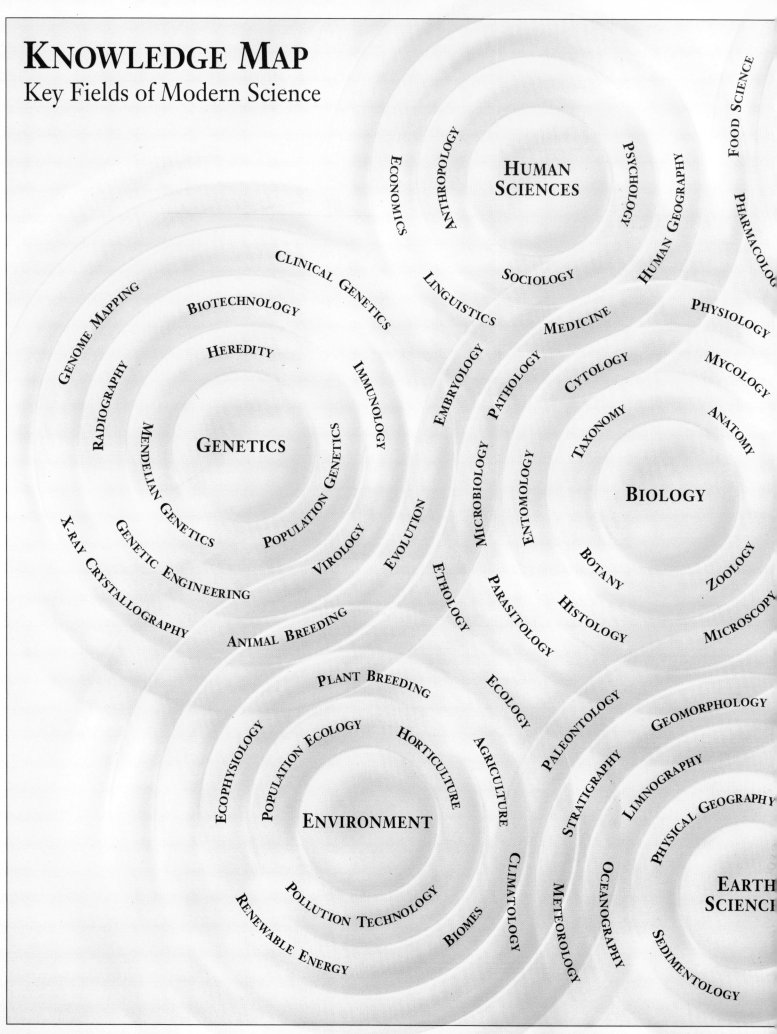

HUMAN SCIENCES

ECONOMICS
ANTHROPOLOGY
PSYCHOLOGY
FOOD SCIENCE
HUMAN GEOGRAPHY
PHARMACOLOGY
SOCIOLOGY
LINGUISTICS
MEDICINE
PHYSIOLOGY
CLINICAL GENETICS
BIOTECHNOLOGY
GENOME MAPPING
HEREDITY
IMMUNOLOGY
EMBRYOLOGY
PATHOLOGY
CYTOLOGY
MYCOLOGY
RADIOGRAPHY
ANATOMY
TAXONOMY
MENDELIAN GENETICS
GENETICS
BIOLOGY
POPULATION GENETICS
MICROBIOLOGY
ENTOMOLOGY
GENETIC ENGINEERING
VIROLOGY
EVOLUTION
BOTANY
ZOOLOGY
X-RAY CRYSTALLOGRAPHY
HISTOLOGY
MICROSCOPY
ANIMAL BREEDING
ETHOLOGY
PARASITOLOGY
PLANT BREEDING
ECOLOGY
PALEONTOLOGY
GEOMORPHOLOGY
ECOPHYSIOLOGY
POPULATION ECOLOGY
HORTICULTURE
AGRICULTURE
STRATIGRAPHY
LIMNOGRAPHY
PHYSICAL GEOGRAPHY
ENVIRONMENT
CLIMATOLOGY
METEOROLOGY
OCEANOGRAPHY
EARTH SCIENCE
POLLUTION TECHNOLOGY
BIOMES
RENEWABLE ENERGY
SEDIMENTOLOGY

CHEMISTRY

Analytical Chemistry
Physical Chemistry
Chromatography
Spectrography
Industrial Chemistry
Electrochemistry
Photochemistry
Materials Science
Astrophysics
Inorganic
Periodic Table
Biochemistry

ASTRONOMY

Stellar Evolution
Stellar Astronomy
Galactic Astronomy
Observational Astronomy
Celestial Mechanics
Space Exploration
Grand Unified Theories
Lunar Studies
Relativity
Cosmology
Radioastronomy
Spectroscopy

PHYSICS

Quantum Physics
Atomic Physics
Radioactivity
Electricity
Particle Physics
Solid State Physics
Mechanics
Nuclear Physics
Biophysics
Heat and Thermodynamics
Statics
Optics
Dynamics
Cryogenics
Acoustics
Gravity
Recording
Electronics
Tectonic Theory
Electromagnetism
Seismology
Telecommunications
Geochemistry
Mineralogy
Planetology
Mapping
Info Technology
CAD Design

MATH AND COMPUTERS

Robotics
Mensuration
Artificial Intelligence
Software Engineering
Multimedia
Number Theory
Logic
Algebra
Statistics
Geometry
Mathematical Analysis
Word Processing
Graphics
Applied Math
Spreadsheets

KNOWLEDGE MAP
Modern Physics

ELECTRONICS

The study of the flow of ELECTRICITY in a vacuum or low-pressure gas (vacuum tubes), in circuit components such as capacitors, inductors and resistors, and in semiconductor devices. The subject includes the use of electronic devices in TELECOMMUNICATIONS equipment, computers and other media.

ATOMIC PHYSICS

The study of the atom and its behavior, explaining physical phenomena in terms of events at the atomic scale. It is also concerned with the internal structure of atoms, although increasingly this subject is assigned to PARTICLE PHYSICS.

PARTICLE PHYSICS

The branch of physics concerned with the study of elementary particles – the fundamental units of the atom. These are divided into hadrons (proton, neutron, pion, etc) and leptons (electron, muon, neutrino etc). Because hadrons themselves consist of quarks, it can be argued that the true elementary particles are quarks and leptons.

ELECTRO-MAGNETISM

The study of phenomena that result from the combination of an electric field and a magnetic field, and how they interact with electric charges (stationary or moving). It is the theoretical basis of devices such as electromagnets, generators and motors, and applies to radio waves, light and other electromagnetic radiation.

QUANTUM PHYSICS

The theory that radiant energy is emitted (or absorbed) only in discrete units of "packets" of energy called quanta. For electromagnetic radiation the quanta are photons. The way in which radiation interacts with matter is the subject of quantum mechanics, which acknowledges the dual wave-particle nature of radiation.

RADIOACTIVITY

Some unstable atomic nuclei spontaneously disintegrate and emit radiation, consisting of one or more types; alpha particles (helium nuclei), beta particles (electrons) or gamma rays (very short-wavelength electromagnetic radiation). With naturally occurring isotopes, the phenomenon is termed natural radioactivity. Artificial radioactivity is produced by bombarding atomic nuclei with neutrons or heavier particles.

RELATIVITY

The physical expression of the idea that observations depend on the observer's viewpoint as much as on what is being observed. Albert Einstein's special theory (1905) states that physical laws remain the same in all frames of reference, but that the speed of light is constant. One conclusion is the equivalence of mass and energy. Einstein's general theory (1915) includes gravitation, and describes the Universe as four-dimensional.

NUCLEAR PHYSICS

The branch of physics that studies the nuclei of atoms and how they react with each other. It is concerned with the generation of energy, either by fission (splitting heavy atoms to yield lighter ones) or fusion joining light atoms to form heavier ones). Fission is the basis of today's nuclear reactors.

CRYOGENICS

The study of the physical properties of substances at very low temperatures. The science depends on methods of liquefying hydrogen and helium, with which temperatures a few degrees above absolute zero can be achieved.

ELECTRICITY

The study of phenomena that result from the presence of electric charges. Stationary charges are the subject of static electricity; moving charges are involved in current electricity. This branch of physics deals with the electrical properties of substances, and with the generation of electricity and its practical applications.

MECHANICS

The branch of physics concerned with the interplay between matter and any forces that act on it. Its divisions STATICS and DYNAMICS can be applied to study the design and workings of simple machines.

DYNAMICS

The branch of physics that studies the way forces produce movement in objects, alternatively called KINETICS. Two key concepts are inertia and momentum (related to mass and velocity), which cannot be varied without the application of an external force.

STATICS

The branch of physics that studies the way forces act on stationary objects, which do not move (there is no change in momentum) because the forces acting on them are balanced; it is a branch of MECHANICS.

SOLID-STATE PHYSICS

The study of the electrical properties of solids, notably semiconductors and superconductors (in terms of their atomic structures). Electronic components consisting entirely of solids, such as semiconductor diodes and transistors, are termed solid-state devices and their uses are part of ELECTRONICS.

HEAT AND THERMODYNAMICS

Heat is a form of energy that can be equated with the kinetic energy (energy of motion) of the vibrating atoms or molecules of a substance at any temperature above absolute zero. At absolute zero, such motion ceases – which is one way of stating the third law of thermodynamics. The other laws are concerned with relationships between forms of energy and their effects in physical systems.

ACOUSTICS

The study of sound. It includes measuring the speed of sound in various media, and how sound waves can be reflected and refracted. Interference – the combination of two or more sound waves – is also important in acoustics.

RECORDING

The manipulation of electrical signals that represent data, sounds or images. Typically the signals are outputs from a microphone or from a video camera. Audio and video recording techniques include magnetic recording (on tape or disk) and recording on laser disks. Magnetic recording is also important for computer memory.

OPTICS

The study of light, its production and detection. Other topics include the determination of the speed of light; how light is reflected and refracted (geometrical optics); and how light waves are diffracted and dispersed (physical optics). Interference, the combination of two or more light waves, is also part of physical optics.

TELE-COMMUNICATIONS

The application of electric or electronic devices to convey audio or video signals or other data over distances. Telecommunications systems include radio, television, telegraphy, telephony, radar and facsimile transmission (fax). The principal "carriers" of information are electric currents (along wires or coaxial cables), radio and microwave signals and light conducted along fiber-optic cables.

TIMECHART

PHYSICS, LIKE ALL SCIENCES, uses both theory and experiment in its attempt to understand nature. One of the earliest discoveries was theoretical: the relation between the pitch of musical notes and the length of vibrating strings. Pythagoras of Samos (c580–500 BC) found that harmonious sounds were given out by strings whose lengths were in simple numerical ratio. From this observation the conviction grew that mathematics could reveal the realities of nature.

Another theoretical idea of the Greeks that has survived was put forward by Democritos (c460–c361 BC): the notion that everything in the Universe consists of minute indivisible particles. He had no evidence for his view. In fact, even in the early 1900s respectable scientists could doubt the existence of atoms. The first reliable estimate of the size of a molecule or combination of atoms was made in 1908 by Jean-Baptiste Perrin (1870–1942).

Another Greek thinker, Aristotle (384–322 BC), tried to interpret the world without using abstractions such as atoms and mathematics. He emphasized observation. He reasoned that

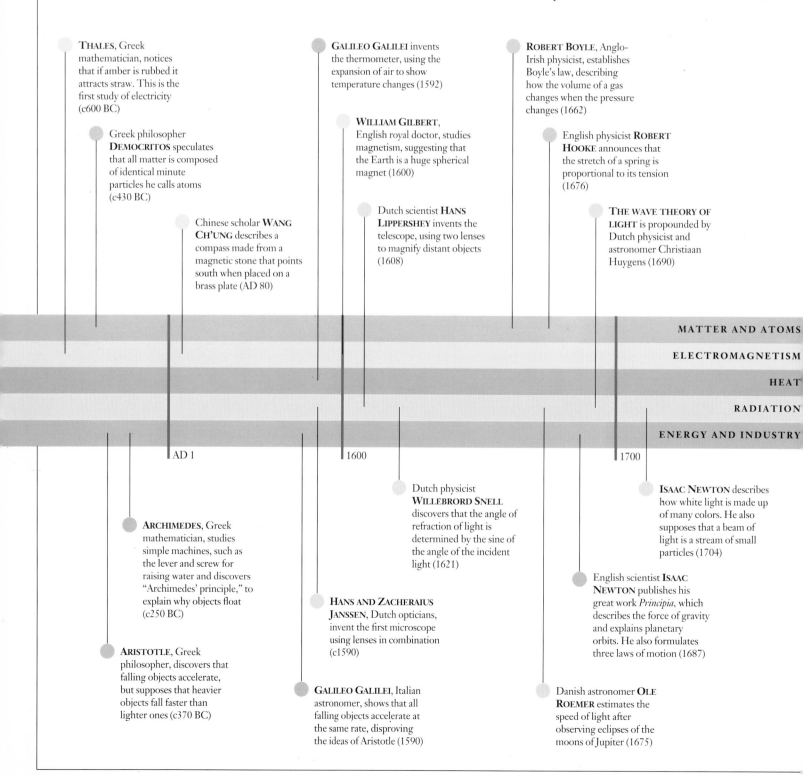

THALES, Greek mathematician, notices that if amber is rubbed it attracts straw. This is the first study of electricity (c600 BC)

Greek philosopher DEMOCRITOS speculates that all matter is composed of identical minute particles he calls atoms (c430 BC)

Chinese scholar WANG CH'UNG describes a compass made from a magnetic stone that points south when placed on a brass plate (AD 80)

GALILEO GALILEI invents the thermometer, using the expansion of air to show temperature changes (1592)

WILLIAM GILBERT, English royal doctor, studies magnetism, suggesting that the Earth is a huge spherical magnet (1600)

Dutch scientist HANS LIPPERSHEY invents the telescope, using two lenses to magnify distant objects (1608)

ROBERT BOYLE, Anglo-Irish physicist, establishes Boyle's law, describing how the volume of a gas changes when the pressure changes (1662)

English physicist ROBERT HOOKE announces that the stretch of a spring is proportional to its tension (1676)

THE WAVE THEORY OF LIGHT is propounded by Dutch physicist and astronomer Christiaan Huygens (1690)

MATTER AND ATOMS

ELECTROMAGNETISM

HEAT

RADIATION

ENERGY AND INDUSTRY

AD 1   1600   1700

ARCHIMEDES, Greek mathematician, studies simple machines, such as the lever and screw for raising water and discovers "Archimedes' principle," to explain why objects float (c250 BC)

ARISTOTLE, Greek philosopher, discovers that falling objects accelerate, but supposes that heavier objects fall faster than lighter ones (c370 BC)

HANS AND ZACHERAIUS JANSSEN, Dutch opticians, invent the first microscope using lenses in combination (c1590)

GALILEO GALILEI, Italian astronomer, shows that all falling objects accelerate at the same rate, disproving the ideas of Aristotle (1590)

Dutch physicist WILLEBRORD SNELL discovers that the angle of refraction of light is determined by the sine of the angle of the incident light (1621)

Danish astronomer OLE ROEMER estimates the speed of light after observing eclipses of the moons of Jupiter (1675)

English scientist ISAAC NEWTON publishes his great work Principia, which describes the force of gravity and explains planetary orbits. He also formulates three laws of motion (1687)

ISAAC NEWTON describes how white light is made up of many colors. He also supposes that a beam of light is a stream of small particles (1704)

matter consisted of four elements – earth, water, air and fire – with a fifth element in the heavens holding the stars in place. Motion occurred when an object sought its rightful place. Unfortunately, Aristotle did not do experiments to test his ideas.

Archimedes (c287–212 BC) introduced the scientific method into physics. He discovered the principles underlying levers and the principle of flotation by measuring the effects that occurred under standard conditions and then deducing general laws. All subsequent advances relied on mathematical interpretation of

experiments. In 1604, for example, Galileo Galilei (1564–1642) performed experiments to test Aristotle's ideas about falling objects. By rolling spheres down a slope, Galileo was able to deduce an exact relationship between the speed of a falling object and the time it took to fall. This work was brought to a conclusion by Isaac Newton (1642–1727). In 1687, Newton published his three laws of motion and the mathematical theory of gravity.

During the 18th and 19th centuries physics advanced on many fronts. Notably, the science of electricity and electromagnetism

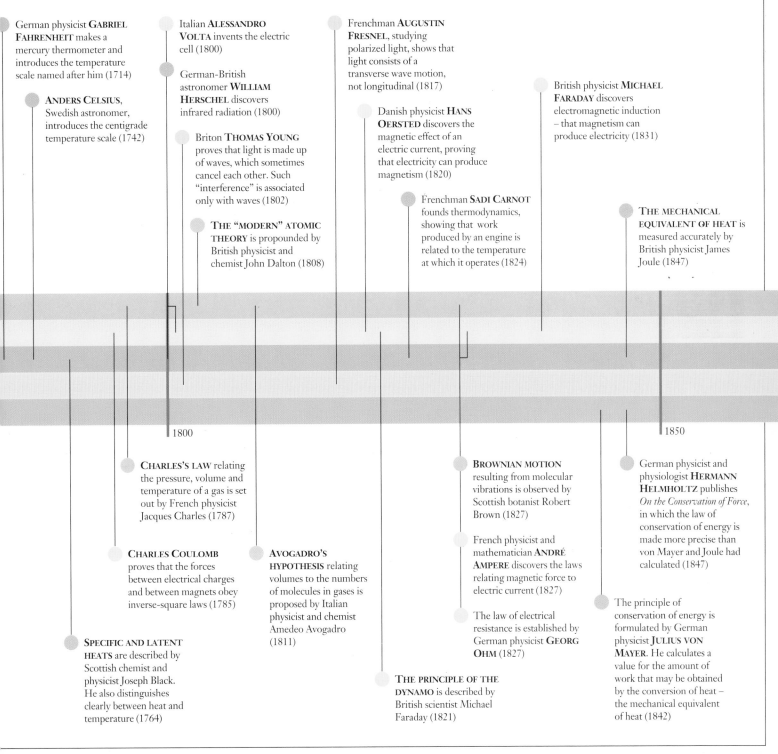

German physicist **GABRIEL FAHRENHEIT** makes a mercury thermometer and introduces the temperature scale named after him (1714)

ANDERS CELSIUS, Swedish astronomer, introduces the centigrade temperature scale (1742)

Italian **ALESSANDRO VOLTA** invents the electric cell (1800)

German-British astronomer **WILLIAM HERSCHEL** discovers infrared radiation (1800)

Briton **THOMAS YOUNG** proves that light is made up of waves, which sometimes cancel each other. Such "interference" is associated only with waves (1802)

THE "MODERN" ATOMIC THEORY is propounded by British physicist and chemist John Dalton (1808)

Frenchman **AUGUSTIN FRESNEL**, studying polarized light, shows that light consists of a transverse wave motion, not longitudinal (1817)

Danish physicist **HANS OERSTED** discovers the magnetic effect of an electric current, proving that electricity can produce magnetism (1820)

Frenchman **SADI CARNOT** founds thermodynamics, showing that work produced by an engine is related to the temperature at which it operates (1824)

British physicist **MICHAEL FARADAY** discovers electromagnetic induction – that magnetism can produce electricity (1831)

THE MECHANICAL EQUIVALENT OF HEAT is measured accurately by British physicist James Joule (1847)

1800

1850

CHARLES'S LAW relating the pressure, volume and temperature of a gas is set out by French physicist Jacques Charles (1787)

CHARLES COULOMB proves that the forces between electrical charges and between magnets obey inverse-square laws (1785)

AVOGADRO'S HYPOTHESIS relating volumes to the numbers of molecules in gases is proposed by Italian physicist and chemist Amedeo Avogadro (1811)

BROWNIAN MOTION resulting from molecular vibrations is observed by Scottish botanist Robert Brown (1827)

French physicist and mathematician **ANDRÉ AMPERE** discovers the laws relating magnetic force to electric current (1827)

The law of electrical resistance is established by German physicist **GEORG OHM** (1827)

German physicist and physiologist **HERMANN HELMHOLTZ** publishes *On the Conservation of Force*, in which the law of conservation of energy is made more precise than von Mayer and Joule had calculated (1847)

SPECIFIC AND LATENT HEATS are described by Scottish chemist and physicist Joseph Black. He also distinguishes clearly between heat and temperature (1764)

THE PRINCIPLE OF THE DYNAMO is described by British scientist Michael Faraday (1821)

The principle of conservation of energy is formulated by German physicist **JULIUS VON MAYER**. He calculates a value for the amount of work that may be obtained by the conversion of heat – the mechanical equivalent of heat (1842)

developed. In 1800, Alessandro Volta (1745–1827) made the first electrical cell (battery). Hans Oersted (1777–1851) discovered the magnetic effect of an electric current in 1820. Michael Faraday (1791–1867) went on to develop the electric motor and generator, or dynamo. The culmination of these developments was the work of James Clerk Maxwell (1831–1879), who showed that electric and magnetic fields could be reproduced as wave motion, and that light was electromagnetic radiation of this sort. This theory seemed to fill the last gap in the Newtonian explanation of the

Universe. All that was required, it seemed, was to calculate the consequences of the theory to greater and greater accuracy.

The Newtonian edifice started to crumble in 1887, when Albert Michelson (1852–1931) and Edward Morley (1838–1923) tested the theory that light waves were carried by an invisible medium called the "ether". Not only did they fail to detect the ether but they also discovered that the velocity of light is constant regardless of the motion of the observer. From this result, Albert Einstein (1879–1955) derived his special theory of relativity in

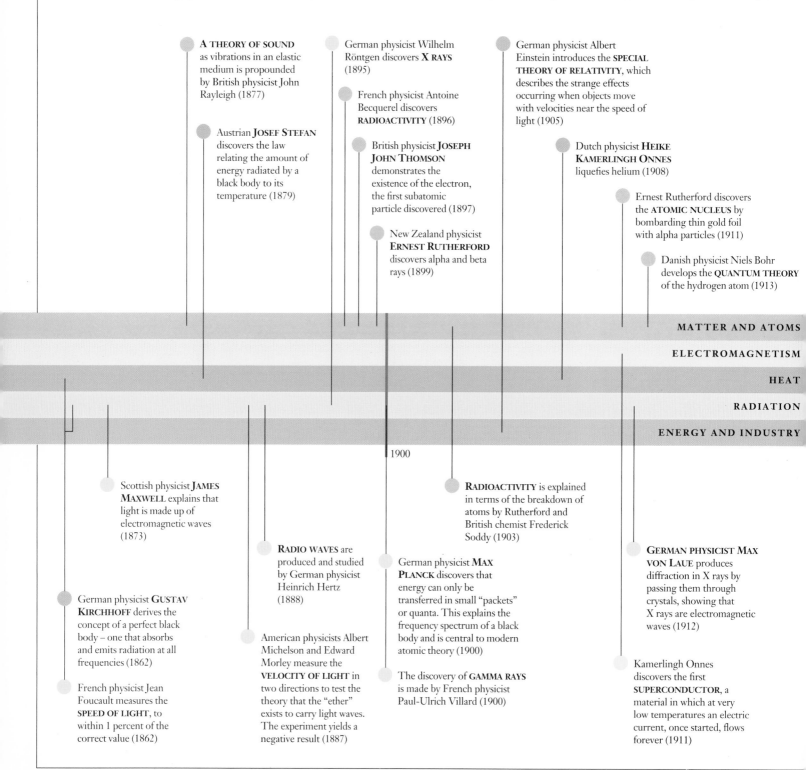

A THEORY OF SOUND as vibrations in an elastic medium is propounded by British physicist John Rayleigh (1877)

Austrian **JOSEF STEFAN** discovers the law relating the amount of energy radiated by a black body to its temperature (1879)

German physicist Wilhelm Röntgen discovers **X RAYS** (1895)

French physicist Antoine Becquerel discovers **RADIOACTIVITY** (1896)

British physicist **JOSEPH JOHN THOMSON** demonstrates the existence of the electron, the first subatomic particle discovered (1897)

New Zealand physicist **ERNEST RUTHERFORD** discovers alpha and beta rays (1899)

German physicist Albert Einstein introduces the **SPECIAL THEORY OF RELATIVITY**, which describes the strange effects occurring when objects move with velocities near the speed of light (1905)

Dutch physicist **HEIKE KAMERLINGH ONNES** liquefies helium (1908)

Ernest Rutherford discovers the **ATOMIC NUCLEUS** by bombarding thin gold foil with alpha particles (1911)

Danish physicist Niels Bohr develops the **QUANTUM THEORY** of the hydrogen atom (1913)

MATTER AND ATOMS

ELECTROMAGNETISM

HEAT

RADIATION

ENERGY AND INDUSTRY

1900

Scottish physicist **JAMES MAXWELL** explains that light is made up of electromagnetic waves (1873)

German physicist **GUSTAV KIRCHHOFF** derives the concept of a perfect black body – one that absorbs and emits radiation at all frequencies (1862)

French physicist Jean Foucault measures the **SPEED OF LIGHT**, to within 1 percent of the correct value (1862)

RADIO WAVES are produced and studied by German physicist Heinrich Hertz (1888)

American physicists Albert Michelson and Edward Morley measure the **VELOCITY OF LIGHT** in two directions to test the theory that the "ether" exists to carry light waves. The experiment yields a negative result (1887)

German physicist **MAX PLANCK** discovers that energy can only be transferred in small "packets" or quanta. This explains the frequency spectrum of a black body and is central to modern atomic theory (1900)

The discovery of **GAMMA RAYS** is made by French physicist Paul-Ulrich Villard (1900)

RADIOACTIVITY is explained in terms of the breakdown of atoms by Rutherford and British chemist Frederick Soddy (1903)

GERMAN PHYSICIST MAX VON LAUE produces diffraction in X rays by passing them through crystals, showing that X rays are electromagnetic waves (1912)

Kamerlingh Onnes discovers the first **SUPERCONDUCTOR**, a material in which at very low temperatures an electric current, once started, flows forever (1911)

1905. His theory meant that Newton's laws needed modification at high velocities. In 1915 Einstein showed how the presence of matter modifies the space around it, producing gravitational effects. This explained an anomaly in the orbit of Mercury, which could not be accounted for by Newton's theory of gravity.

The early part of the 20th century also saw the development of atomic theory. The discovery of the electron in 1897 showed that there were particles smaller than atoms. Also, atoms were now seen to be destructible; in 1903 Ernest Rutherford (1871–1937)

explained that radioactivity is due to the breakdown of atoms. His discovery of the nucleus showed something of the internal structure of the atom. Niels Bohr (1885–1962), Louis de Broglie (1892–1987), Erwin Schrödinger (1887–1961) and Werner Heisenberg (1901–1976) developed the quantum theory of atomic structure in which the position and momentum of an electron cannot be known without a degree of uncertainty. This seems to imply that "reality" cannot be known precisely. Yet quantum theory has proved the most accurate theory of all time.

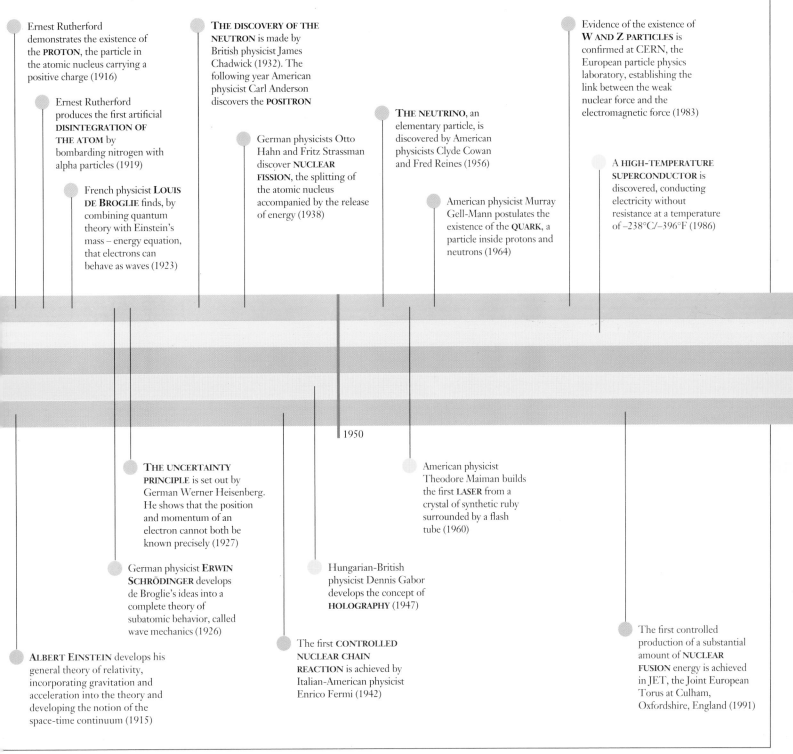

Ernest Rutherford demonstrates the existence of the **PROTON**, the particle in the atomic nucleus carrying a positive charge (1916)

Ernest Rutherford produces the first artificial **DISINTEGRATION OF THE ATOM** by bombarding nitrogen with alpha particles (1919)

French physicist **LOUIS DE BROGLIE** finds, by combining quantum theory with Einstein's mass – energy equation, that electrons can behave as waves (1923)

THE DISCOVERY OF THE NEUTRON is made by British physicist James Chadwick (1932). The following year American physicist Carl Anderson discovers the **POSITRON**

German physicists Otto Hahn and Fritz Strassman discover **NUCLEAR FISSION**, the splitting of the atomic nucleus accompanied by the release of energy (1938)

THE NEUTRINO, an elementary particle, is discovered by American physicists Clyde Cowan and Fred Reines (1956)

American physicist Murray Gell-Mann postulates the existence of the **QUARK**, a particle inside protons and neutrons (1964)

Evidence of the existence of **W AND Z PARTICLES** is confirmed at CERN, the European particle physics laboratory, establishing the link between the weak nuclear force and the electromagnetic force (1983)

A HIGH-TEMPERATURE SUPERCONDUCTOR is discovered, conducting electricity without resistance at a temperature of $-238°C/-396°F$ (1986)

1950

THE UNCERTAINTY PRINCIPLE is set out by German Werner Heisenberg. He shows that the position and momentum of an electron cannot both be known precisely (1927)

German physicist **ERWIN SCHRÖDINGER** develops de Broglie's ideas into a complete theory of subatomic behavior, called wave mechanics (1926)

ALBERT EINSTEIN develops his general theory of relativity, incorporating gravitation and acceleration into the theory and developing the notion of the space-time continuum (1915)

Hungarian-British physicist Dennis Gabor develops the concept of **HOLOGRAPHY** (1947)

The first **CONTROLLED NUCLEAR CHAIN REACTION** is achieved by Italian-American physicist Enrico Fermi (1942)

American physicist Theodore Maiman builds the first **LASER** from a crystal of synthetic ruby surrounded by a flash tube (1960)

The first controlled production of a substantial amount of **NUCLEAR FUSION** energy is achieved in JET, the Joint European Torus at Culham, Oxfordshire, England (1991)

Physics KEYWORDS

aberration

Any defect that causes an optical instrument to produce a distorted image. Aberration occurs because of tiny variations in lenses and mirrors, and because different wavelengths of light are reflected or refracted by different amounts. *See* **chromatic aberration, spherical aberration** and **astigmatism.**

absolute temperature

A temperature measured with respect to absolute zero – that is, measured on the Kelvin temperature scale. There can be no negative values in absolute temperatures.

absolute zero

The lowest theoretically possible temperature, zero kelvin (0 K), which is equivalent to -273.15°C (-459.67°F), at which molecules are motionless. Although the third law of thermodynamics states that it is impossible to reach absolute zero exactly, a temperature of 2×10^{-9} K (two billionths of a degree above absolute zero) was produced in 1989 by Finnish scientists.

CONNECTIONS
HEAT ENERGY **68**
HEATING AND COOLING **70**

absorption

The phenomenon in which material retains radiation of particular wavelengths, such as a piece of blue glass absorbing all visible light except the wavelengths in the blue part of the spectrum. It also refers to the partial loss of energy resulting from light and other electromagnetic waves passing through a medium. In nuclear physics, absorption is the capture by elements such as boron of neutrons produced by fission in a reactor. *See* **moderator.**

acceleration

The rate of change of the velocity of a moving object measured in units such as meters per second per second (m/s^2) or feet per second per second (ft/s^2). Velocity is a vector quantity (it possesses both magnitude and direction). For this reason an object traveling at constant speed is said to be accelerating if its direction of motion changes. The average acceleration a of an object traveling in a straight line over a period of time t may be calculated using the formula: acceleration a = change of velocity/t or, where u is initial velocity and v final velocity: $a = (v - u)/t$. A negative value for velocity shows that the object is slowing down (decelerating). Acceleration due to gravity (the acceleration of free fall) is the acceleration of an object falling freely under the influence of the Earth's gravitational field.

CONNECTIONS
MATTER ON THE MOVE **60**
FORCE OF GRAVITY **62**
MECHANICAL ENERGY **64**

accelerator

A device used for bringing charged subatomic particles (such as protons and electrons) up to high speeds and energies. When such high-energy particles collide, the fragments formed provide information about the fundamental forces of nature. The first accelerators used high voltages to generate a strong, unvarying electric field. Charged particles were accelerated as they passed through the electric field. However, because the voltage produced by a generator is limited, these accelerators were replaced by machines in which the particles passed through regions of alternating electric field, receiving a succession of small pushes to accelerate them. The first of these accelerators was the linear accelerator or linac, which consists of a line of metal tubes through which the particles travel. The particles are accelerated by electric fields in the gaps between the tubes. Another way of making use of an electric field is to bend the path of a particle into a circle so that it passes repeatedly through the same electric field. Early accelerators directed the particle beam onto a stationary target; large modern accelerators usually collide beams of particles that are traveling in opposite directions.

accumulator

Another name for a **secondary cell.**

adiabatic

A process that occurs without loss or gain of heat, especially the expansion or contraction of a gas in which a change takes place in the pressure, volume and temperature, though no heat is allowed to enter or leave.

aerodynamics
The branch of fluid physics that studies the forces exerted by air or other gases in motion, particularly the airflow around objects (such as land vehicles, bullets, rockets and aircraft) moving at speed through the atmosphere. For maximum efficiency, the aim is usually to design the shape of an object to produce a **streamline flow**, with a minimum of turbulence in the moving air.

airfoil
Part of an aircraft shaped so that it creates lift or controls direction of flight as the plane flies through the air. The best-known example is an airplane's wing, which is shaped in such a way that air flows farther, and hence faster, across the upper surface (than the lower one), resulting in lower air pressure on top of the wing and hence upward lift. The airfoil (wing) is also designed to minimize drag in the direction of motion.

alpha particle
A positively charged, high-energy particle emitted from the nucleus of a radioactive atom. It is a product of the spontaneous disintegration of radioactive elements (*see* **radioactivity**) and is identical with the nucleus of a helium atom: that is, it consists of two protons and two neutrons. Alpha decay, the process of emission, transforms one element

AIRFOIL

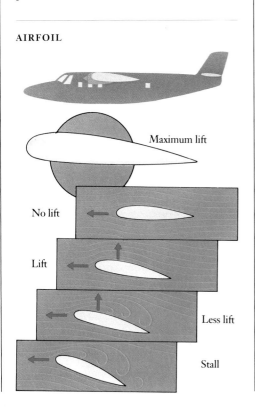

Maximum lift

No lift

Lift

Less lift

Stall

into another, decreasing the atomic number by two and the atomic mass by four. Alpha particles have a range of only a few centimeters in air and can be stopped by a sheet of paper. *See also* **beta particle**.

alternating current (AC)
Electric current that flows in alternately reversed directions around a circuit. In a power station, electricity is usually generated as alternating current. The advantage of alternating current over direct current, as from a battery, is that its **voltage** can be raised or lowered economically by means of a **transformer.**

alternator
An electricity generator that produces an **alternating current (AC)**. It consists of a coil or coils that rotate in the magnetic field produced by one or more permanent magnets or electromagnets supplied by an independent direct current source. The frequency of the alternating current depends on the speed at which the coils rotate and the number of magnetic poles.

ammeter
An instrument that measures the electric current, usually in amperes, in an electric circuit. A common type is the moving-coil meter, which measures direct current, but can, in the presence of a rectifier, also measure alternating current. Other types include the moving iron ammeter (DC) and the thermoammeter (AC).

amorphous
The term given to any non-crystalline solid – that is, one in which there is no long-range order in its lattice. Amorphous metals have good electrical and thermal conductivities and other metal-like qualities; however, unlike crystalline metal solids, their atomic arrangements are not periodically ordered. Metallic glasses are amorphous materials.

Ampere's law
The law that gives the magnetic induction at a point caused by an electric current in terms of the size of the current and the length of the conductor carrying it.

amplifier
An electronic device or circuit that magnifies an input current or voltage, increasing its amplitude to produce a gain.

amplitude
The maximum displacement of an oscillation from the equilibrium position. For a wave motion, it is the height of a crest (or the depth of a trough). With a sound wave, for

example, amplitude corresponds to the intensity (loudness) of the sound. *See* **wave** and **amplitude modulation**.

amplitude modulation (AM)
A method by which radio waves are altered for broadcasting them. Amplitude modulated waves are constant in frequency, but the amplitude of the continuously transmitted radio carrier wave varies in accordance with the signal being broadcast. *See also* **frequency modulation (FM)**.

analog
(Of a quantity or device) Changing continuously; a digital quantity or device varies in a series of distinct steps. For example, an analog clock measures time by means of a continuous movement of hands around a dial, whereas a digital clock measures time with a numerical display that changes in a series of individual steps.

analog-to-digital converter (ADC)
An electronic circuit that converts an analog signal into a digital one. Such a circuit is used to convert the signal from an analog device into a digital signal for input into a computer. For example, many sensors designed to measure physical quantities, such as temperature and pressure, produce an analog signal in the form of a continuously varying voltage. This must be passed through an ADC before computer input and processing.

angle of incidence
The angle between a ray of light striking a surface (such as a lens or mirror) and a line at 90° (known as the normal) to that surface.

angle of reflection
When light is reflected by a mirror, the angle between the reflected ray and the normal (line at right angles to the mirror).

angle of refraction
When light passing from one transparent medium to another is refracted (bent), the angle between the refracted ray and the normal (line at right angles to the surface between the two media). *See* **critical angle**.

angular momentum
The product of the moment of inertia and the angular velocity of an orbiting or rotating object (*see* **momentum**). The angular momentum of an object of mass m traveling at a velocity v in a circular orbit of radius R is expressed as mvR. Angular momentum is conserved, and if any of the values alter (such as the radius of orbit), the other values (such as the velocity) will also change to maintain the value of angular momentum.

anode

A positive electrode. In an electron gun it attracts electrons (from the cathode) through a hole and therefore electrons flow out of the device from the anode. In this case the anode is made positive by means of an external source. However, in a battery the anode is the electrode that spontaneously becomes positive as a result of a chemical reaction. It therefore attracts electrons to it from an external circuit.

antiparticle

A subatomic particle corresponding in mass and other properties to another particle but with the opposite electrical charge, magnetic properties, or coupling to other fundamental forces. For example, an electron carries a negative charge whereas its antiparticle, the positron, carries a positive one. When a particle and its antiparticle collide, they destroy each other, in the process called annihilation. A substance consisting entirely of antiparticles is known as antimatter.

CONNECTIONS

PROPERTIES OF MATTER **48**
SUBATOMIC PARTICLES **130**

Archimedes' principle

The law stating that an object totally or partly submerged in a fluid experiences an upthrust equal to the weight of fluid it displaces. The upthrust acts vertically through the center of gravity of the displaced fluid. If the weight of the object is less than the upthrust exerted by the fluid, it will float partly or completely above the surface; if its weight is equal to the upthrust, the object will come to equilibrium below the surface; if its weight is more than the upthrust, it will sink.

armature

The rotating part of a DC electric motor or generator.

astigmatism

A defect that occurs in lenses, including that in the eye. It results when the curvature of the lens differs in two perpendicular planes, so that rays in one plane may be in focus while rays in the other are not. With astigmatic eyesight, the vertical and horizontal cannot be in focus at the same time; correction is by the use of a cylindrical lens that reduces the overall focal length of one plane so that both planes are seen in sharp focus.

astronomical telescope

Any telescope that is specifically designed to collect, detect and record electromagnetic radiation from a cosmic source. Optical telescopes can be divided into two types. In refracting telescopes a converging lens is used to collect light and the resulting image is magnified by an eyepiece. By contrast, a reflecting telescope uses a concave mirror to collect and focus the light and a secondary mirror set at an angle of 45 degrees to the main beam to reflect the light to a magnifying eyepiece. *See also* **radio telescope**.

atmospheric pressure

The pressure exerted on the surface of the Earth by the weight of the air above it in the atmosphere. It is measured by a barometer.

atom

The smallest unit of matter that can take part in a chemical reaction and cannot be broken down chemically into anything simpler. An atom is made up of positively charged protons and neutrons (which are not charged) in a central nucleus surrounded by negatively charged electrons. Equal numbers of protons and electrons balance the electrical charge. The atoms of the various elements differ in atomic number, relative atomic mass and chemical behavior. There are 106 different types of atom, corresponding with the 106 known elements as listed in the Periodic Table of the elements. Atoms are much too small to be seen by even the most powerful optical microscope and they are in constant motion.

CONNECTIONS

PROPERTIES OF MATTER **48**
INSIDE THE ATOM **128**
SUBATOMIC PARTICLES **130**
THE UNSTABLE ATOM **132**
NUCLEAR FISSION **134**
NUCLEAR FUSION **136**

atomic clock

A timekeeping device regulated by various periodic processes occurring in atoms and molecules, such as atomic vibration or the frequency of absorbed or emitted radiation. Modern atomic clocks use the cesium atom, which produces or absorbs radiation of a very precise frequency (9,192,631,770 Hz) that varies by less than one part in 10 billion. This frequency has been used in the SI units definition of the second.

atomic number

The number of protons in the nucleus of an atom. Each chemical element has its own atomic number, which gives that element its identity. The atomic number is also known as proton number.

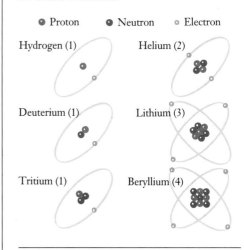

ATOMIC NUMBER

- Proton   ● Neutron   ○ Electron

Hydrogen (1)   Helium (2)

Deuterium (1)   Lithium (3)

Tritium (1)   Beryllium (4)

atomic spectrum

The spectrum produced by the absorption or emission of photons as electrons move between separate energy states in an atom. Different atoms have their own characteristic spectrum.

atomic weight

Another name for **relative atomic mass**.

audio disk

A type of phonograph record, originally made of thermoplastic resin, in which a reproducing needle (pickup) follows tiny undulations in a spiral groove while the record is rotated at a constant speed.

Avogadro constant

The number of atoms or molecules in one mole of a substance, equal to 6.02253×10^{23}. Also known as Avogadro's number.

Avogadro's law

The law that states that equal volumes of all gases at the same temperature and pressure contain the same number of molecules.

background radiation

Ever-present radiation from natural sources, such as outer space and radioactive minerals on Earth.

ballistics

The study of the motion and impact of projectiles, especially bullets, bombs and missiles. For projectiles from a gun, exterior factors include temperature, barometric pressure and wind strength; and for nuclear missiles these include the speed at which the Earth turns on its axis.

barometer

An instrument for measuring atmospheric pressure. Most commonly used are the mercury barometer and the aneroid barometer. A simple mercury barometer consists of a column of mercury in a glass tube, roughly 0.75 m (2.5 ft) high (closed at one end, curved upward at the other), which is balanced by the pressure of the atmosphere on the open end. Any change in pressure is reflected in a change in the height of the column. In an aneroid barometer, a shallow cylindrical corrugated metal box containing a partial vacuum expands or contracts in response to changes in pressure.

battery

An energy-storage device that allows the release of electricity on demand, usually made up of one or more electrical cells. Primary-cell batteries are disposable; secondary-cell batteries, or accumulators, are rechargeable. Primary-cell batteries are an extremely uneconomical form of energy, because they produce only 2 percent of the power used in their manufacture. The common dry cell is an example of a primary-cell battery. The lead-acid car battery is a secondary-cell battery which is continually recharged by the car's dynamo or alternator. It consists of sets of lead (positive) and lead dioxide (negative) plates in an electrolyte of sulfuric acid (battery acid). Rechargeable nickel-cadmium (NiCad) batteries are used in portable electronic devices as a stable, corrosion-free short-term source of power.

CONNECTIONS

ELECTRIC CURRENT 80

PRODUCING ELECTRIC CURRENT 82

ELECTROLYSIS 88

---

beats

Volume variations that occur when two tones of nearly equal frequency sound at the same time. They are caused by interference.

becquerel

Unit of radioactivity in the SI system, equal to the number of atoms (of a radioactive element) that disintegrate in one second.

beta particle

An electron or positron ejected from the nucleus of a radioactive atom undergoing spontaneous disintegration. Beta particles do not exist in the nucleus but are created on disintegration (beta decay), when a neutron converts to a proton with the emission of an electron. Beta particles are more penetrating than alpha particles, but less so than gamma radiation; they can travel several meters in air, but they are stopped by 2–3 mm of aluminum. *See also* **alpha particle**.

bioluminescence

The production of light by living organisms. Such light is usually produced by the oxidation of luciferin, a reaction catalyzed by the enzyme luciferase. Bioluminescence is a feature of many deep-sea fish, crustaceans and other marine animals. On land, bioluminescence occurs in some nocturnal insects such as glow-worms and fireflies, and in certain bacteria and fungi.

black body

A hypothetical object in space that completely absorbs all thermal (heat) radiation striking it and is a perfect emitter of thermal radiation. A practical approximation can be made by using a small hole in the wall of a constant-temperature enclosure. The radiation emitted by a black body is of all wavelengths, but with maximum radiation at a particular wavelength that depends on the object's temperature. As the temperature increases, the wavelength of maximum intensity becomes shorter.

Bohr theory

Theory of atomic structure that pictures atoms as being made up of a positively charged central nucleus surrounded by orbiting electrons (located in **orbitals**).

boiling point

The temperature at which the application of more heat to a liquid can raise its temperature no further, and the liquid is converted into vapor. At the boiling point the saturated vapor pressure of a liquid equals the pressure of the atmosphere, and thus the boiling point varies with pressure and with altitude. The lower the pressure, the lower the boiling point and vice versa. The boiling point of water under standard pressure (at sea level) is 100°C (212°F).

Boyle's law

One of the gas laws stating that the volume (V) of a given mass of gas at a constant temperature is inversely proportional to its pressure (p) – that is, pV = constant. For example, if the pressure of a gas doubles, its volume will be reduced by a half. The law was discovered in 1662 by the Anglo-Irish physicist and chemist Robert Boyle.

breeder reactor

A type of nuclear reactor that produces more fissile material than it consumes in operation. *See* **nuclear reactor**.

Brownian motion

The continuous random motion of particles in a fluid medium (gas or liquid) as they are subjected to impact from the molecules of

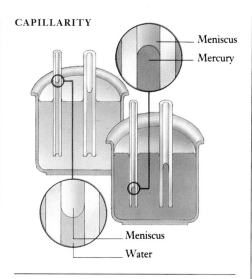

Meniscus
Mercury
Meniscus
Water

---

the medium. The smaller the particle, the more extensive the motion. Brownian motion provides evidence for the **kinetic theory** of matter.

bubble chamber

A device for observing the nature and movement of atomic particles, and their interaction with radiations. A bubble chamber consists of a container filled with a superheated liquid through which ionizing particles move and collide. The paths of these particles are revealed by strings of bubbles, which can be photographed and studied. *See also* **cloud chamber**.

calorie

The amount of heat needed to raise the temperature of one gram of water by one degree Celsius. It equals 4.184 joules (a joule is the SI unit of heat).

capacitance

The electrical property of a **capacitor** that determines how much charge can be stored in it for a given potential difference between its plates. It is equal to the ratio of the stored electrical charge (in coulombs) to the potential difference (in volts). The SI unit of capacitance is the farad (F).

capacitor

A device for storing electric charge in electronic circuits. In its simplest form it consists of two or more conducting plates separated by an insulating layer called a **dielectric**, which may be vacuum, paper, mica, or titanium dioxide. *See* **capacitance**.

capillarity

The spontaneous movement of liquids up or down fine-bore tubes due to unbalanced molecular attraction at the boundary between the liquid and the tube. If the liquid molecules near the boundary are more strongly attracted to molecules in the material of the tube than to the liquid molecules nearby, the liquid will rise. If the liquid molecules are less attracted to the material of the tube than to other liquid molecules, the liquid will fall.

carburetor

A device in a gasoline engine that produces a combustible mixture of volatile fuel and air by mixing them in controlled proportions. Fuel is discharged through jets into an air stream under the pressure difference created by the velocity of air as it flows through a nozzle or jet.

Carnot cycle

An ideal heat-engine cycle of maximum efficiency, devised in the early 19th century as an attempt to improve the efficiency of the steam engine. The Carnot cycle shows that the thermal efficiency of such an engine depends on the temperatures at which it accepts and discards heat. It is made up of a series of changes in the physical condition of the gas in the heat engine in the following order: **1** isothermal expansion (without change of temperature), **2** adiabatic expansion (without change of heat content), **3** isothermal compression and **4** adiabatic compression. The principles derived from a study of the Carnot cycle are important in the fundamentals of heat and the laws of **thermodynamics**.

carrier wave

An electromagnetic wave of specified frequency and amplitude that is emitted by a radio transmitter in order to carry a broadcast signal. The information is superimposed onto the carrier wave by adjusting its amplitude or frequency. *See* **amplitude modulation** and **frequency modulation**.

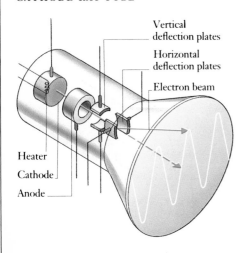

CATHODE-RAY TUBE

Vertical deflection plates
Horizontal deflection plates
Electron beam
Heater
Cathode
Anode

cathode

A negative **electrode**. In an electron gun, electrons are emitted by the cathode and attracted to the anode. Such a cathode may be cold, with emission being due to electric fields or photoemission, but is usually hot, meaning that the cathode is heated. In a battery the cathode is the electrode that spontaneously becomes negative during discharge and from which electrons flow onto an external circuit. *See also* **anode**, **battery** and **cell**.

cathode-ray tube

A device in which a well-defined and controllable beam of electrons (cathode rays) is produced from an **electron gun**. The electrons are directed onto a phosphor-coated screen to produce a visible display. Such tubes are used as picture tubes in television sets, in radar sets and in a computer display.

CONNECTIONS

ELECTRONICS AND SEMICONDUCTORS **92**

RADAR AT WORK **122**

TELEVISION CAMERA **124**

TELEVISION RECEIVER **126**

---

cell

A device in which chemical energy is converted into electrical energy; in other words, a battery. A cell contains two **electrodes** immersed in an electrolyte in a container. A spontaneous chemical reaction in the cell generates a negative charge (an excess of electrons) on one electrode and a positive charge on the other. This causes a current to flow in any external circuit. *See* **battery**, **primary cell** and **secondary cell**.

Celsius

A scale of temperature, previously called centigrade, in which the range from freezing to boiling of water is divided into 100 degrees, with freezing point at 0°C and boiling point at 100°C. The degree centigrade (°C) was officially renamed Celsius in 1948 in order to avoid confusion with the angular measure known as the centigrade (representing one hundredth of a grade). The Celsius scale is named for the Swedish astronomer Anders Celsius, who invented it in 1742.

center of curvature

The center of a sphere of which a lens surface or a curved mirror forms a part. *See* **radius of curvature**.

center of gravity

The point in an object around which its weight is evenly balanced. In a uniform grav-

itational field, the center of gravity is the same as the center of mass.

center of mass

The point in an object from which its total weight appears to originate and can be assumed to act. A symmetrical homogeneous object such as a sphere has its center of mass at its physical center; a hollow shape (such as a cup) may have its center of mass in space inside the hollow. For an object to be in stable equilibrium, a perpendicular line down through its center of mass must run within the boundaries of its base; if tilted until this line falls outside the base, the object becomes unstable and topples over.

centrifugal force

An apparent (but not real) inertial force that acts radially outward from a spinning or orbiting object, thus balancing the **centripetal force** (which is real). For an object of mass m moving with a velocity v in a circle of radius r, the centrifugal force F is equal to mv^2/r (outward).

centripetal force

The force that acts radially inward on an object moving in a curved path. For example, with a weight whirled in a circle at the end of a length of string, the centripetal force is the tension in the string. For an object of mass m moving with a velocity v in a circle of radius r, the centripetal force F equals mv^2/r (inward). It is equal and opposite to an apparent inertial force known as the **centrifugal force**, which is directed away from the center of curvature of the path.

chain reaction

A fission reaction in nuclear physics that is maintained because neutrons released by the splitting of some atomic nuclei themselves go on to split others, releasing even more neutrons; and so on as long as the reaction is maintained. Thus, one nucleus of the isotope uranium-235 can disintegrate with the production of two or three neutrons, which cause similar fission of adjacent nuclei. These in turn produce more neutrons. If the total amount of material exceeds a critical mass, the chain reaction may cause a nuclear explosion. Such a chain reaction can be controlled (as in a **nuclear reactor**) by absorbing excess neutrons. A chain reaction may also occur in chemical reactions. *See also* **nuclear fission**.

CONNECTIONS

THE UNSTABLE ATOM **132**

NUCLEAR FISSION **134**

---

charge-coupled device (CCD)

A semiconductor device made up of layers of silicon that release electrons when struck by incoming light; it is used for forming images electronically. The electrons are stored in pixels and read off into a computer at the end of the exposure. CCDs have almost entirely replaced photographic film for applications such as astrophotography in which extreme sensitivity to light is necessary.

Charles' law

One of the **gas laws** stating that the volume (V) of a given mass of gas at standard pressure is directly proportional to its absolute temperature (T) in Kelvin. The formula for Charles' law is given as TV = constant. The law was discovered by the French physicist Jacques Charles in 1787, but more accurately established by the French chemist Joseph Gay-Lussac in 1802. The gas increases by $1/273$ of its volume at 0°C for each 1°C rise of temperature.

chromatic aberration

The type of **aberration** in which the image is surrounded by colored fringes, because light of different colors is brought to different focal points by a lens.

cloud chamber

A device used for tracking ionized particles. It consists of a vessel fitted with a piston and filled with air or other gas, supersaturated with water vapor. When the volume of the vessel is suddenly expanded by moving the piston outward, the vapor cools and a cloud of tiny droplets forms on any nuclei, dust or ions present. As fast-moving ionizing particles collide with the air or gas molecules, they show as visible tracks. *See also* **bubble chamber**.

coaxial cable

Any electric cable that consists of a solid or stranded central conductor insulated from and surrounded by a solid or braided conducting tube or sheath. The central and outer conductor have the same axis.

coherent light

Light in which two or more sets of waves have a constant phase relationship – that is, the peaks and troughs of all the waves are in step. **Lasers** produce coherent light.

cohesion

The kinetic interaction between molecules of a liquid that enables them to form thin films and drops (*see* **surface tension**). In gases the molecules are separated by relatively larger distances so that less cohesion occurs. *See also* **Joule-Kelvin effect.**

colloid

A mixture of two or more substances in which small particles are dispersed in a gas (an aerosol) or in a liquid (a gel).

color

The property of light emitted or reflected from an object depending on its wavelength. Visible white light consists of electromagnetic radiation which, if refracted through a prism, can be spread out into a spectrum, in which the various colors correspond to different wavelengths. The colors are red, orange, yellow, green, blue, indigo and violet. Light entering the human eye is either reflected from the objects we see or emitted by hot or luminous objects. Sources of light have a characteristic spectrum or range of wavelengths. Hot gases, such as the vapor of sodium in street lights, emit light at particular wavelengths. The pattern of these wavelengths is unique to each gas, and can be used to identify it.

When an object is illuminated by white light, some of the wavelengths are absorbed and some are reflected to the eye of an observer. The object appears colored because of the mixture of wavelengths in the reflected light. For instance, a red object absorbs all wavelengths falling on it except those in the red end of the spectrum. This process, known as subtraction, also explains why certain mixtures of paints produce different colors. Blue and yellow paints when mixed together produce green because between them the yellow and blue pigments absorb all wavelengths except those around green. A combination of three pigments – cyan (blue-green), magenta (blue-red), and yellow – can produce any color when mixed.

It is possible to produce any color by mixing the three primary colors. This process is called color mixing by addition, and is used to produce the color on a television screen where glowing phosphor dots of red, green and blue combine. Pairs of colors that produce white light, such as orange and blue, are called complementary colors.

COLOR

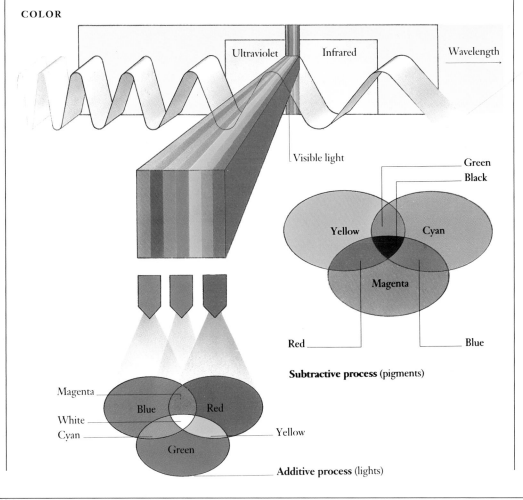

Subtractive process (pigments)

Additive process (lights)

CONNECTIONS

LIGHT AND THE SPECTRUM **106**
PRODUCING LIGHT **108**
DISPERSION AND DIFFRACTION **114**
LIGHT FROM LASERS **116**
TELEVISION CAMERA **124**
TELEVISION RECEIVER **126**

commutator
Device that reverses the direction of an electric current, as in a DC electric motor.

compact disc
A digitally encoded metal disk, read by a laser. Compact discs (CDs), about 12 cm (4.5 in) across, are commonly used in domestic hi-fi systems for high-quality sound reproduction. They are made of aluminum with a transparent plastic coating; the metal disk is etched underneath (by a laser beam) with microscopic pits that carry the digital code representing the sounds. When the disc is played back, a laser beam reads the code and produces signals that are changed into near-exact replicas of the original sounds. Compact discs can also be used to store text or pictures. *See also* **audio disk.**

compass
An instrument for finding direction. The most commonly used is a magnetic compass, consisting of a needle of magnetic material with the north-seeking pole marked, freely pivoted at its center and mounted on a base on which the points of the compass are indicated. With correct use, the north-seeking pole points to magnetic north, from which true north (at the end of the Earth's axis) can be found from tables of magnetic corrections. Compasses that are not dependent on a magnet – gyrocompasses and radiocompasses – are unaffected by iron and by anomalies of the Earth's magnetic field, and are widely used in ships and aircraft.

compression ratio
The ratio of the total volume enclosed in the cylinder of an internal-combustion engine at the beginning of the compression stroke to the volume at the end of the compression stroke. For gasoline engines the ratio is in the range 8.5–9 to 1; for diesel engines it is in the range 12–18 to 1.

Compton effect
The change in energy (and therefore wavelength) of X-ray photons when they strike electrons, which increase in energy.

conduction
Thermal conduction is the direct transmission of heat through a substance from a region of high temperature to a region of low temperature. In gases and most liquids, this is achieved by collisions between atoms and molecules with those possessing lower kinetic energy. In solids and liquid metals, it occurs mainly through the migration of fast-moving electrons. Electrical conduction is the passage of electric charge under the influence of an electric field.

conductor
Any material that conducts heat or electricity (as distinguished from an insulator or non-conductor). A good conductor has a high electrical or heat conductivity, and is usually rich in free electrons. A poor conductor (such as glass or porcelain) has few free electrons. Metals are good conductors. Carbon is non-metallic and yet (in some of its forms) a relatively good conductor of heat and electricity. Substances such as silicon and germanium, with intermediate conductivities that are improved by heat, light or an electric field, are known as semiconductors. *See* **insulator** and **semiconductor**.

conservation of energy
A basic principle of physics which states that energy in a closed system can be neither created nor destroyed but remains constant.

conservation of momentum
A principle of physics which states that, for two colliding objects, their total momentum before impact is the same as their total momentum after impact.

control rod
A neutron-absorbing rod that can be pushed in and out of the core of a **nuclear reactor** to control a chain reaction.

convection
The movement of heat energy through a liquid or gas which expands as its temperature rises; the expanded material, being less dense, rises above colder and therefore denser material. Convection currents are used to carry hot water in pipes in some heating systems.

cosmic rays
Extremely short wavelength radiation that originates in outer space. It forms part of background radiation.

Coulomb's law
The law that states that the force between two charged particles Q_1 and Q_2 a distance d apart is proportional to the product of the charges and inversely proportional to the square of the distance between them. The law is usually written as $F = Q_1Q_2/4\pi\varepsilon_0d^2$ where ε_0 is the absolute permittivity of free space.

couple
A pair of forces acting on an object that are equal in magnitude and opposite in direction, but do not act along the same straight line. The two forces produce a turning effect that tends to rotate the object; however, no single resultant (unbalanced) force is produced and so the object does not move from one place to another.

critical angle
For a light ray passing from a denser to a less dense medium (such as from glass to air), the critical angle is the smallest angle of incidence at which the emergent ray grazes the surface of the denser medium – at an angle of refraction of 90 degrees. When the angle of incidence is greater than the critical angle, the ray is reflected back (total internal reflection) and does not pass out of the denser medium; when the angle of incidence is less than the critical angle, the ray passes into the less dense medium. *See also* **refraction**.

critical mass
The minimum mass of fissile material that can sustain a continuous **chain reaction**. Below this mass, too many neutrons escape from the surface for a chain reaction to continue; above it, the reaction may accelerate into a nuclear explosion, unless controlled by a **moderator** (as in a **nuclear reactor**).

critical pressure
The pressure at which a gas may be liquefied at its **critical temperature**.

critical temperature
The temperature above which a gas cannot be liquefied by an increase of pressure. In electronic engineering it is the temperature at which magnetic materials lose their magnetic properties.

CRITICAL ANGLE

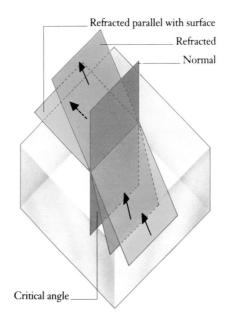

Refracted parallel with surface
Refracted
Normal
Critical angle

cryogenics

The study of very low temperatures (approaching **absolute zero**), including their production and the exploitation of special properties associated with them, such as the disappearance of electrical resistance (**superconductivity**). Low temperatures can be produced by the **Joule-Kelvin** (or Joule-Thompson) **effect**. Gases such as oxygen, hydrogen and helium may be liquefied in this way, and temperatures of 0.3 K can be reached.

Further cooling requires magnetic methods; a magnetic material, in contact with the substance to be cooled and with liquid helium, is magnetized by a strong magnetic field. The heat generated by the process is carried away by the helium. When the material is then demagnetized, its temperature falls; temperatures of around 10^{-3} K have been achieved in this way. A similar process, called nuclear adiabatic expansion, was used to produce the lowest temperature so far recorded: 2×10^{-9} K.

At temperatures near absolute zero, materials can display unusual properties. Some metals, such as mercury and lead, exhibit superconductivity. Liquid helium loses its viscosity and becomes a "superfluid" when cooled to below 2 K; in this state it flows up the sides of its container. Cryogenics has several industrial applications. Electronic components called Josephson junctions, which could be used in very fast computers, need low temperatures to function. Some magnetic levitation (**maglev**) systems must be maintained at low temperatures.

crystal

A homogeneous solid substance with an orderly three-dimensional arrangement of its atoms or molecules, thereby creating an external surface of clearly defined smooth faces. All crystals of the same substance have the same characteristic angles between their faces, although they may not have the same appearance because the faces can grow at different rates. The external form of the crystal is known as the crystal habit. Crystals fall into a number of crystal systems or groups, classified on the basis of the relationship of three or four imaginary axes that intersect at the center of any perfect, undistorted crystal. Examples of crystals include minerals such as quartz, and chemicals such as inorganic salts.

crystallography

The scientific study of the form and structure of crystals. In 1912 it was found that the shape and size of the regular repeating atomic patterns (unit cells) in a crystal could be determined by passing X-rays through a sample. This method, known as X-ray diffraction, opened up a new way of observing atomic structure. It has been found that many substances have a unit cell that exhibits all the symmetry of the whole crystal; in table salt, for instance, the unit cell is an exact cube.

crystalloid

A substance that is not dispersed as a **colloid**.

current

See **electric current**.

Curie temperature

See **critical temperature**.

cyclotron

A type of accelerator that accelerates atomic particles to high speeds by making them follow a spiral path between the poles of two D-shaped magnets.

cylinder

The tubular chamber in which the piston of an engine or pump moves back and forth. The inside diameter of the cylinder is called the bore, and the distance traveled by the piston inside the cylinder is the stroke.

Dalton's law of partial pressure

In a mixture of gases, each of the component gases exerts the pressure it would exert if it alone occupied the total volume.

Daniell cell

A primary cell that has a zinc cathode and a copper anode, both dipping into an electrolyte of dilute sulfuric acid.

de Broglie wavelength

The wavelength associated with a fast-moving particle, equal to **Planck's constant** divided by the particle's momentum (the product of its mass and velocity).

decay

The spontaneous transformation of one radioactive nucleus into a daughter nucleus, which may or may not be radioactive, with the emission of one or more particles or photons. The time required for half the original nuclei to decay is called the half-life. The same terms are also applied to elementary particles that spontaneously transform into other particles. *See* **alpha particle, beta particle** and **gamma rays**.

decibel (dB)

The unit used originally to compare sound intensities and subsequently electrical or electronic power outputs, now also used to compare voltages. An increase of 10 dB is equivalent to a tenfold increase in intensity or power, and a 20-fold increase in voltage. A whisper has a sound intensity of about 20 dB; 140 dB (a jet aircraft taking off nearby) is the threshold of pain.

degree

The unit of temperature difference, usually defined as a certain fraction of the fundamental temperature scale interval. For most thermometers the fundamental interval is the difference in temperature between the freezing point and **boiling point** of water. *See* **Celsius, Fahrenheit** and **Kelvin scale**.

density

A measure of the compactness of a substance. The density of a substance is defined as its mass per unit volume and is measured in units such as kilograms per cubic meter (kg/m^3) or pounds per cubic foot (lb/ft^3). The density D of a mass m occupying a volume V is given by the formula: $D = m/V$. Relative density is the ratio of the density of a substance to that of water at $4°C$.

depression of freezing point

The lowering of the freezing point of a liquid (solvent) when a solid (solute) is dissolved in it which, for dilute solutions of a non-volatile solvent at constant pressure, is proportional to the concentration of solute (whose relative molecular mass can thereby be calculated).

dew point

The temperature at which air becomes saturated with water vapor. Below this temperature, water condenses as dew or mist.

diamagnetism

The weak magnetization of a material in the opposite direction to an applied magnetic field. Although all substances are diamagnetic, it may be masked by other, stronger forms. *See* **magnetism** and **paramagnetism**.

dielectric

A substance (an insulator such as ceramic, rubber or glass) that is capable of supporting different **electric charges** in different regions. Dielectrics are often used in **capacitors** and to reduce very strong electric fields by polarization, producing a field to oppose the original field. The strength of the original field is thus reduced by a factor called the dielectric constant. The dielectric constant is also called relative permittivity.

diesel engine

A type of internal-combustion engine that burns a lightweight fuel oil. A diesel engine operates by compressing air until it becomes sufficiently hot to ignite the fuel. It is a piston-in-cylinder engine, like a gasoline engine, but only air (rather than an air-and-fuel mixture) is taken into the cylinder on the first piston stroke (down). The piston moves up and compresses the air in order to heat it to a very high temperature. The fuel oil is then injected into the hot air, where it ignites (without the need for a spark plug), driving the piston down on its power stroke. The principle of the diesel engine was first explained in England by Herbert Akroyd in 1890, and was applied practically in an engine built by Rudolf Diesel in Germany in 1892.

diffraction

The spreading of a wave motion (such as light or sound) as it passes an object and expands into a region beyond the geometric shadow of the object. This accounts for interference phenomena observed at the edges of opaque objects or discontinuities between different media in the path of a wave train. The phenomenon of diffraction causes the slight spreading of light into colored bands at the shadow of a straight edge and also accounts for the ability of sound to carry around corners.

CONNECTIONS

SOUND ENERGY **96**

LIGHT AND THE SPECTRUM **106**

REFRACTION AND LENSES **112**

DISPERSION AND DIFFRACTION **114**

---

diffraction grating

An optical device for producing spectra. In its simplest form it consists of a plate of glass or metal, ruled with close, equidistant parallel lines, used for separating a wave train such as a beam of incident light into its component wavelengths. The spectra are produced by diffraction effects from the lines, which act as a very large number of equally spaced parallel slits, from 70 to 1800 lines per millimeter.

diffusion

The spontaneous movement of matter in which molecules or ions mix through normal thermal agitation. The rate of diffusion increases with temperature. Migration of ions may also be directed and accelerated by electric fields. Dissolved materials may diffuse both through the material in which they are dissolved and through membranes.

digital

(of a quantity or device) Changing in a series of distinct steps; by contrast, an **analog** quantity or device varies continuously. For example, a digital clock measures time with a numerical display that changes in a series of steps, whereas an analog clock measures time by means of a continuous movement of hands around a dial.

diode

A thermionic tube (valve) with a cold **anode** and a heated **cathode** (or the semiconductor equivalent). A diode allows the passage of direct current only in one direction because a negative potential applied to the anode repels the electrons. Diodes are commonly used in a rectifier to convert **alternating current (AC)** to **direct current (DC)**.

dip

The angle that indicates the direction of the Earth's magnetic field at a point, measured vertically downward from the horizontal.

dipole

A pair of equal and opposite charges located apart, as in some ionic molecules, constitutes an electric dipole. The product of either charge and the distance between them is the dipole moment. A bar magnet or a coil carrying a steady current produces a magnetic dipole. Dipole is also the name given to a simple radio antenna consisting of two metal rods arranged in line with each other.

direct current (DC)

Electric current that flows in one direction, and does not reverse its flow as alternating current (AC) does. The electricity produced by a battery is direct current.

discharge tube

A device in which a gas conducting an electric current emits visible light. It is usually an evacuated glass tube, containing only traces of gas, with electrodes at each end. When a high-voltage current is passed between the electrodes, the few remaining gas atoms in the tube (or some deliberately introduced ones) ionize and emit colored light as they conduct the current along the tube. The light originates as electrons change energy levels in the ionized atoms. By coating the inside of the tube with a **phosphor**, invisible emitted radiation (such as ultraviolet light) can produce visible light; this is the principle of fluorescent lighting.

dispersion

The separation of electromagnetic radiation into components of different wavelengths. Dispersion is a property of the medium in

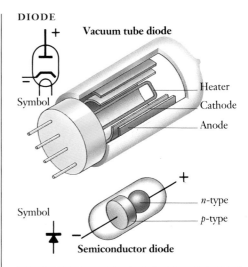

DIODE

Vacuum tube diode

Symbol

Heater
Cathode
Anode

Symbol

n-type
p-type

Semiconductor diode

---

which the wave is propagated. One example is the splitting of white light into a **spectrum** of different colors when it passes through a prism or a **diffraction grating**. Dispersion occurs because the prism (or grating) bends each component wavelength to a slightly different extent. The natural dispersion of light through raindrops creates a rainbow.

distance ratio

The distance moved by the input force (effort) divided by the distance moved by the output force (load) in any machine. The ratio indicates the movement magnification achieved and is equivalent to the machine's velocity ratio.

domain

A small area in a magnetic material that behaves like a tiny magnet. The **magnetism** of the material is due to the movement of electrons in the atoms of the domain. In an unmagnetized sample of material, the domains point in random directions, or form closed loops, so that there is no overall magnetization. In a magnetized sample, the domains line up so that their magnetic effects combine to produce a strong overall magnetism.

doping

The addition of small quantities of impurities to a **semiconductor** material in order to achieve the desired properties in diodes, transistors and other electronic components.

Doppler effect

The apparent change in the observed frequency (or wavelength) of waves due to relative motion between the wave source and the observer. The Doppler effect is responsible for the perceived change in pitch of a siren as it approaches and recedes, and for the red shift of light from distant stars. It is named for the Austrian physicist Christian Doppler.

dry battery

A battery composed of dry cells – that is, **primary cells** or **secondary cells** in which the contents are in the form of paste. Many flashlight, radio and general purpose batteries are examples of Leclanché cells, a form of dry cell in which a central carbon electrode is immersed in an electrolyte of a manganese dioxide and ammonium chloride paste. The zinc container forms the negative electrode (with an outer plastic or paper wrapper).

dynamics

The mathematical and physical study of changing motion of objects acted on by forces. Also called kinetics. *See also* **statics**.

dynamo

A simple generator or machine for transforming mechanical energy into electrical energy. A basic dynamo consists of a powerful field magnet with a suitable conductor, usually in the form of a coil (armature), rotating between its poles. The mechanical energy of rotation is converted into an electric current in the armature.

> ### CONNECTIONS
>
> ELECTRICITY GENERATION **84**
>
> ELECTRIC MOTORS **86**

echo

The return of a sound wave, or of a radar or sonar signal, by reflection from an object.

echogram

The record produced by an **echo sounder**. When charting the depth of the sea bed the information can be permanently displayed on a trace-type recorder. In less expensive instruments the depth may be instantly output on a dial or digital-type display.

ELASTICITY

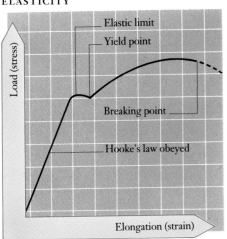

echo sounder

An instrument consisting of a transmitter that emits ultrasonic pulses and a receiver that detects objects under water by means of sonar (using reflected high-frequency sound waves). By measuring the time taken for an echo to return to the transmitter and by knowing the speed of a sonar signal, it is possible to calculate the range of the object causing the echo (echolocation).

efficiency

A measure of mechanical performance. Efficiency is defined as the useful work output (work done by the machine) divided by the work input (work put into the machine), usually expressed as a percentage. Losses of energy caused by friction mean that efficiency is always less than 100 percent, although it can be very high for an electrical machine with no moving parts, such as a transformer.

> ### CONNECTIONS
>
> SIMPLE MACHINES **66**
>
> MEASURING AND USING HEAT **72**
>
> ELECTRICITY GENERATION **84**

effort

Any applied force acting against inertia, equivalent to the input into any machine. *See* **distance ratio** and **force ratio**.

elastic limit

When a deforming force is applied to an object, the object's elastic limit is the point beyond which it does not return to its original shape or dimensions after the deforming force is removed. Up to this point any deformation is elastic; beyond it, deformation is permanent. *See* **Hooke's law, elasticity** and **yield point**.

elasticity

The ability of a solid to recover its shape once deforming forces are removed. Elastic materials include metals, some plastics and rubber; however, all materials have some degree of elasticity. *See* **yield point**.

electric arc

A continuous electric discharge of high current between two electrodes which gives out brilliant light and heat. The phenomenon is exploited in the carbon-arc lamp, once widely used in film projectors, and in the electric-arc furnace, where an arc struck between very large carbon electrodes and the metal charge provides the heating. In arc welding, an electric arc provides the heat to fuse the metals being joined.

ELECTRIC CHARGE

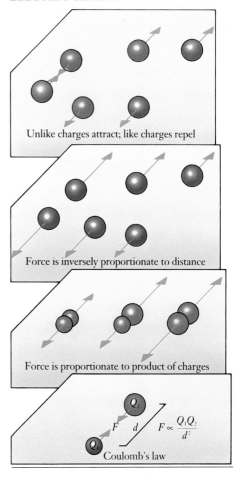

Unlike charges attract; like charges repel

Force is inversely proportionate to distance

Force is proportionate to product of charges

$$F \propto \frac{Q_1 Q_2}{d^2}$$

Coulomb's law

electric charge

A fundamental property of matter and the basis of all electrical phenomena, including electricity. Electric charge may be either positive or negative. In atoms, electrons possess a negative charge, and protons possess an equal positive charge. An object that has equal quantities of positive and negative charge is said to be neutral. Two objects that have the same charge repel each other, whereas objects with opposite or unlike charges attract each other, because each is in the electric field of the other. The SI unit of measurement for electric charge is the coulomb (represented by the symbol C). A flow of charge (such as electrons through a copper wire) constitutes an electric current; the flow of current is measured in amperes (represented by the symbol A).

> ### CONNECTIONS
>
> MAGNETS AND FIELDS **76**
>
> STATIC ELECTRICITY **78**
>
> ELECTRIC CURRENT **80**
>
> PRODUCING ELECTRIC CURRENT **82**
>
> ELECTRICITY GENERATION **84**
>
> ELECTRICITY AND OTHER ENERGY **90**

electric current

The flow of electrically charged particles through a conducting circuit due to a potential difference. The current at any point in a circuit is the amount of charge flowing per second; its SI unit is the ampere (coulombs per second). When current flows in a component that possesses resistance, electrical energy is converted into heat energy. If the resistance of the component is R ohms and the current through it is I amperes, then the heat energy W (in joules) generated in a time t seconds is given by the formula $W = I^2Rt$.

electric field

The region around an electric charge in which a charged object is subject to a force.

electric furnace

A type of furnace that utilizes the extreme temperatures produced in an **electric arc**.

electric motor

A machine that converts electrical energy to mechanical energy. A simple direct-current (DC) motor consists of a horseshoe-shaped permanent magnet with a wire-wound coil (armature) mounted so that it can rotate between the poles of the magnet. A commutator reverses the current (from a battery) fed to the coil on each half-turn, which rotates because of the mechanical force exerted on a conductor carrying a current in a magnetic field. Other types of electric motor include the induction motor (which does not require a commutator) and the linear induction motor, which produces linear rather than rotary motion.

electric potential

The energy required to bring a unit charge from infinity (zero potential) to the point in an electric field at which the potential is specified. Its unit is the volt. The potential difference is the difference in value of the electric potential at the two points, and is equivalent to the work done in moving unit charge from one point to the other.

electrode

A terminal through which an electric current passes in or out of a conducting substance, such as the anode or cathode in a battery.

ELECTROMAGNETIC WAVES

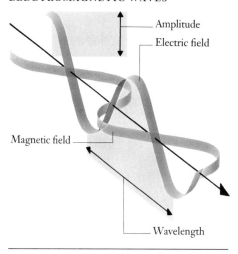

- Amplitude
- Electric field
- Magnetic field
- Wavelength

The terminals that emit and collect the flow of electrons in thermionic tubes (valves) are also called electrodes.

electrolysis

The chemical changes produced by passing an electric current through a solution or molten salt (the electrolyte), resulting in the migration positive ions (cations) to the negative electrode (cathode) and negative ions (anions) to the positive electrode (anode). Electrolysis is used in industrial processes as well as in electrochemical analysis.

electrolyte

A solution or molten material that conducts electric current, due to the presence of **ions**.

electromagnet

A soft iron core surrounded by a coil of wire, which acts as a magnet when an electric current flows through it. Electromagnets have many applications, as in switches, electric bells, solenoids and metal-lifting cranes.

electromagnetic induction

The production of an electromagnetic force in a conductor when it moves in a magnetic field, as in a dynamo.

electromagnetic radiation

Energy resulting from moving electric charges and traveling with oscillating electric and magnetic fields at right angles to each other. *See also* **electromagnetic spectrum**.

electromagnetic spectrum

The range of wavelengths over which electromagnetic waves extend. The longest (10^5 to 10^{-3} m) are radio waves; next longest are infrared waves, followed by the narrow band of visible light (4-7×10^{-7} m), ultraviolet waves (10^{-7} to 10^{-9} m), X rays (10^{-9} to 10^{-11} m) and gamma rays (10^{-11} to 10^{-14} m).

electromagnetic waves

Oscillating electric and magnetic fields traveling through space at the speed of light. The (limitless) range of possible wavelengths or frequencies of electromagnetic waves make up the **electromagnetic spectrum**.

electromagnetism

The phenomenon by which magnetic fields can be produced by the flow of electrons in an electric current. *See* **electromagnet**.

electromotive force

The greatest voltage that can be generated by a particular source of electric current. When the source is connected in a circuit, some of the energy it supplies is lost as it drives current across its own internal resistance, and so its terminal voltage (the potential difference across its terminals) is less than its electromotive force.

electron

A stable, negatively charged elementary particle which is a constituent of all atoms, and a member of the class of particles known as leptons. The electrons in each atom surround the nucleus in groupings called shells; in a neutral atom the number of electrons is equal to the number of protons in the nucleus. The shell structure is responsible for the chemical properties of the atom. Electrons are the basic particles of electricity. Each carries a charge of 1.602192×10^{-19} coulombs. All electric charges are multiples of this quantity.

electron gun

A device for producing a steady narrow beam of electrons. It usually consists of a heated cathode, a control grid and one or more annular anodes inserted in an evacuated glass tube. The electrons emitted by the cathode are attracted to the final anode, which has a hole through which they pass. An electron gun is used as the source of electrons in **cathode-ray tubes** and **electron microscopes**.

electronics

The design and application of devices that amplify and switch electrically, without moving parts. The first device was the thermionic tube (using electrons moving in a vacuum), now superseded by solid-state technology; the transistor (invented in 1949); and multitransistor circuits in the form of silicon chips, for reasons of economy, compactness, reliability and low power consumption. Based on the thermionic tube, electronics developed during the first half of the 20th century, producing radio, radar, tube (valve) television, and the earliest true computers. Solid-state devices can be used to construct extremely complex circuits. There has been rapid development from the simple but distortion-prone techniques of analog signal processing to the superior digital ones, such as from the representation of a sound wave by a voltage varying instant-by-instant in step with the air pressure variation, to a representation in terms of binary numbers. Digital techniques have more obvious uses in calculators, computers and clocks.

electron microscope

An instrument for producing a magnified image by using a beam of electrons from an electron gun instead of light rays. The beam is focused and controlled by an arrangement of electromagnetic coils (magnetic lenses). The electrons form the image on a fluorescent screen or a photographic plate. The wavelength of the electron beam is much shorter than that of light, allowing much greater magnification and resolution to be achieved. A transmission electron microscope passes the beam through a very thin slice of a specimen; a scanning electron microscope produces exterior images.

electroscope

A device for detecting electric charge. A gold-leaf electroscope consists of a vertical conducting rod ending in a pair of rectangular gold leaves, mounted inside an insulated metal case. An electric charge applied to the end of the metal rod causes the gold leaves to diverge, because they each receive a similar charge and so repel each other. The polarity of the charge can be found by bringing up another charge of known polarity and applying it to the metal rod. A like charge has no effect on the gold leaves, whereas an opposite charge neutralizes the charge on the leaves and causes them to collapse.

electrostatic field

An electric field surrounding a stationary electric charge.

electrostatic induction

The occurrence and movement of charges in an electrically conducting material caused by the proximity of charges in another object. It also refers to the separation of charges in a dielectric by an electric field.

elementary particle

Also called fundamental particle; a particle that is a component of atoms and all matter. The most familiar are the **electron, proton** and **neutron**. More than 200 particles have now been identified and categorized into several classes as characterized by their mass, electric charge, spin, magnetic moment and interaction. Although many particles were thought to be nondivisible and permanent, most are now known to be combinations of a small number of basic particles.

elevation of boiling point

The raising of the boiling point of a liquid (solvent) when a solid (solute) is dissolved in it which, for dilute solutions of a non-volatile solvent, is proportional to the concentration of solute (whose relative molecular mass can thereby be calculated).

energy

The capacity for doing work. **Potential energy** derives from position; thus a stretched spring has elastic potential energy, and an object raised to a height above the Earth's surface, or the water in an elevated reservoir, has gravitational potential energy. A lump of coal and a tank of gasoline, together with the oxygen needed for their combustion, have chemical energy. Other sorts of energy include electrical and nuclear energy, heat, light and sound. Moving objects possess **kinetic energy**. Energy can be converted from one form to another, but the total quantity stays the same (in accordance with the conservation of energy principle). For example, as an apple falls, it loses gravitational potential energy but gains kinetic energy. Although energy is never lost, after a number of conversions it tends to finish up as the kinetic energy of random motion of molecules (of the air, for example) at relatively low temperatures. This is "degraded"

EQUILIBRIUM

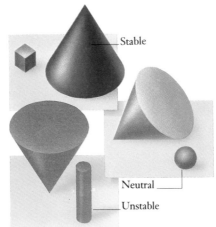

Stable

Neutral

Unstable

energy in that it is difficult to convert it back to other forms. Einstein's special theory of **relativity** correlates any gain in energy, E, with a gain in mass, m, by the equation $E = mc^2$, in which c is the speed of light.

energy level

A defined fixed quantity of energy that a molecule, atom, electron or nucleus can have. An atom, for example, has a fixed energy corresponding to the **orbitals** in which its electrons move around the nucleus. The atom can accept a **quantum** of energy to become an excited atom (*see* **excitation**) if that extra energy will raise an electron to an unoccupied orbital.

entropy

The state of disorder of a system at the atomic, ionic or molecular level; the greater the disorder, the higher the entropy, and the less the ability to do work. For instance, the molecules of water vapor have higher entropy than those of liquid water, which in turn have higher entropy than the molecules in solid crystalline ice. At **absolute zero**, when all molecular motion ceases and order is assumed to be complete, entropy is zero.

equilibrium

A condition in which the forces acting on a particle or system of particles (an object) cancel out, or in which energy is distributed evenly among the particles of a system; alternatively, equilibrium can be the state in which an object is at rest or moving at constant velocity. An object is in thermal equilibrium with its surroundings if no heat enters or leaves it, so that all its parts are at the same temperature as the surroundings. *See also* **center of mass**.

equivalent weight

The mass of a chemical element or compound that can combine with or displace one gram of hydrogen (or 8 grams of oxygen).

escape velocity

The velocity needed by an object to escape from a gravitational field, so that its kinetic energy exceeds its potential energy resulting from gravity. For a rocket escaping from Earth's gravity it is 40,320 km/h.

etchant

A chemical used to remove copper from laminates during the production of printed circuits. *See also* **photoresist process**.

evaporation

A process in which a liquid turns to a vapor without its temperature reaching **boiling point**. Any liquid left to stand evaporates eventually because, at any time, a proportion of its molecules has enough kinetic energy to escape through the intermolecular forces of attraction at the liquid surface into the atmosphere. The rate of evaporation rises with increased temperature because, as the mean kinetic energy of the liquid's molecules rises, so do the number possessing enough energy to escape.

excitation

The addition of energy to a system (such as an atom or a nucleus) that raises it to an energy level above the ground state. The difference between the excited state and the ground state is called the excitation energy. *See* **energy level** and **orbital**.

expansion

The increase in size of a constant mass of substance caused by, for example, increasing its temperature (thermal expansion) or its internal pressure. The expansivity, or coefficient of thermal expansion, of a material is its expansion (per unit volume, area, or length) per degree rise in temperature.

Fahrenheit

The temperature scale in which the freezing point of water is 32°F and the boiling point 212°F; the fundamental interval is 180 degrees. Fahrenheit has been widely replaced by the **Celsius** and **Kelvin** scales.

Faraday constant

The electric charge carried by 1 mole of single-charged ions, equal to 9.6487×10^4 coulombs per mole. It was discovered by the English scientist Michael Faraday.

Faraday's laws of electrolysis

During **electrolysis**, **1** chemical decomposition is proportional to the electric current and **2** the amounts of substances liberated at the electrodes are proportional to their equivalent weights.

Faraday's laws of induction

1 Whenever the magnetic field linking a circuit changes, an electromotive force (e.m.f.) is induced in the circuit. **2** The magnitude of the induced e.m.f. is proportional to the rate of change of the magnetic flux linking the circuit.

fast neutron

A high-energy neutron produced by nuclear fission (which is too fast to produce further fission and sustain a **chain reaction**).

ferromagnetism

A phenomenon in which certain materials become strongly magnetized when placed in a magnetic field and retain their magnetism once the applied field is removed. Ferromagnetic materials, used to make permanent magnets, include iron, cobalt, nickel and their alloys. Ferromagnetism is caused by the forced alignment of **domains** by the external magnetic field.

fiber optics

Technology based on the use of ultra-pure glass fibers having a central core of higher refractive index than the outer cladding and capable of conducting modulated light signals by total internal reflection (*see* **critical angle**). Optical fiber cables can carry much more information (such as telephone calls) more accurately than an equivalent ordinary electric cable.

field

A region in which an object experiences a force as the result of the presence of some other object. A field is a method of representing the way in which objects are able to influence each other. *See* **lines of force**.

fission

See **nuclear fission**.

flotation, law of

The law stating that a floating object displaces its own weight of the fluid in which it floats. It provides an explanation of how an object as large and as heavy as a steel ship can float: the hollow steel hull of the ship sinks into the water until the weight of the water it has displaced is as great as its own weight. The upthrust from the water will then equal the ship's weight and the ship will float. *See* **Archimedes' principle**.

fluid

Any substance in which the molecules are relatively mobile and can flow. Both gases and liquids are considered fluids. A fluid differs from a solid in that it cannot offer any permanent resistance to a change in shape.

fluorescence

The very short-lived emission of light from an object when its atoms are excited by means other than raising its temperature. When exposed to an external source of energy such as ultraviolet light, the outer electrons in atoms of a fluorescent substance

"jump" to a higher energy level (*see* **excitation**). When they return to their former energy level, they emit excess energy as light.

CONNECTIONS

---

fluorescent screen

A screen coated by a fluorescent substance (phosphor) that emits light when excited by an external source of energy, such as X rays or a beam of electrons. Such screens are used in television sets and cathode-ray tubes.

focal length

The distance from the center of a lens or curved mirror to the focal point. For a concave mirror or convex lens, it is the distance at which parallel rays of light are brought to a focus to form a real image. For a convex mirror or concave lens, it is the distance from the center to the point at which a virtual image is formed. The more powerful the lens, the shorter its focal length.

force

Any influence that changes an object's state of rest or its uniform motion in a straight line. Force results in the acceleration of an object in the direction of action of the force; or, if the object is unable to move freely, it may result in its deformation. Force is a vector quantity, possessing both magnitude and direction; its SI unit is the newton.

force ratio

The number of times the load moved by a machine is greater than the effort applied to that machine. In equation terms: force ratio = load/effort. Because it is a ratio, it has no

FORCE RATIO

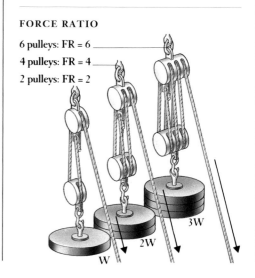

6 pulleys: FR = 6
4 pulleys: FR = 4
2 pulleys: FR = 2

3W
2W
W

units. Force multipliers have a force ratio greater than 1; distance multipliers, or velocity multipliers, have a force ratio less than 1. The exact value of a working machine's force ratio is always less than its predicted value because there will always be some frictional resistance that increases the effort. Also known as mechanical advantage.

four-stroke cycle
The operating cycle of most internal-combustion engines. The "stroke" is an upward or downward movement of a piston in a cylinder. In a gasoline engine the cycle begins with the induction of a fuel mixture as the piston goes down on its first stroke. On the second stroke (up) the piston compresses the mixture in the top of the cylinder. An electric spark then ignites the mixture, and the gases produced force the piston down on its third, power, stroke. On the fourth stroke (up) the piston expels the burned gases from the cylinder into the exhaust.

free fall
The movement of an object caused solely by a gravitational force. Neglecting air resistance or in a vacuum, all objects freely falling under **gravity** have the same (constant) acceleration – the acceleration of free fall.

freezing point
The temperature at which a liquid solidifies. It is the same as the melting point of a solid. The freezing point of water is used to define the lower fixed point on the Celsius and Fahrenheit temperature scales.

frequency
The number of periodic oscillations, vibrations or waves occurring per unit of time. The SI unit of frequency is the hertz (Hz), one hertz being equivalent to one cycle per second. Also known as periodicity.

frequency modulation (FM)
A method by which radio waves are transmitted over long distances by adding a signal to a carrier wave. The frequency of the carrier wave increases when the signal amplitude increases, and decreases when the signal amplitude falls. Because frequency modulated systems are concerned only with changes in the frequency of the carrier wave, they are not affected by the many forms of interference that change the amplitude of the carrier wave. *See also* **amplitude modulation (AM)**.

friction
Resistance to the relative motion of two objects in contact. The frictional force opposing the motion is equal to the applied moving force up to a value known as the limiting friction. If the moving force is less than this value, the objects will not move relative to each other. Any increase in the moving force beyond the limiting friction then causes slippage. The coefficient of friction is the ratio of the force required to achieve this relative motion to the force pressing the two objects together. Friction is greatly reduced by the use of lubricants. In other instances friction is deliberately increased by making the surfaces rough, as in brake linings.

fuel cell
A cell that converts chemical energy directly to electrical energy. It works on the same principle as a **battery** but is continually fed with fuel, usually hydrogen. Fuel cells are more efficient converters of chemical energy than heat engines. Their use to power electrical vehicles is being actively explored.

fundamental forces
The four fundamental interactions that occur between objects even when they are not in physical contact. Together they account for all of the observed forces that occur in the physical universe. There are two long-range forces: gravity, which keeps the planets in orbit around the Sun, and acts between all objects that have mass; and the electromagnetic force, which stops solids from falling apart, and acts between all objects with electric charge. There are two very short-range forces: the weak force, responsible for radioactive decay and other subatomic reactions; and the strong force, which binds together protons and neutrons in the nuclei of atoms. Physicists are working on theories to unify the fundamental forces in a Grand Unified Theory.

fundamental vibration
The standing wave of the longest wavelength that can be established in a vibrating object such as a stretched string or air column. The sound produced by the fundamental vibration is the lowest-pitched (usually dominant) note heard. The fundamental vibration of a string has a stationary node at each end and a single antinode at the center where the amplitude of vibration is greatest. *See* **harmonic**.

fusion
See **nuclear fusion**.

galvanometer
An instrument for detecting small electric currents by their magnetic effect. In a moving-coil galvanometer, a pivoted coil of insulated copper wire surrounds a fixed soft-iron core between the poles of a permanent magnet. When current flows in the coil, a torque is produced by the interaction of the magnetic field and the coil. The strength of the torque can be used to measure the strength of the current. Digital instruments are replacing moving-coil galvanometers.

gamma radiation
Very-high-frequency electromagnetic radiation, similar in nature to **X rays** but of shorter wavelength, emitted by the nuclei of radioactive substances during decay or by the interactions of high-energy electrons with matter. The emission of gamma radiation reduces the energy of the source nucleus, but has no effect on its proton or nucleon numbers. Gamma rays are stopped only by direct collision with an atom and are therefore very penetrating. They are not deflected by magnetic or electric fields.

gas
A form of matter, such as air, in which the molecules move randomly in otherwise empty space, filling any size or shape of container into which it is put. Gases can be liquefied by cooling, which lowers the speed of the molecules and enables attractive forces between them to bind them together.

CONNECTIONS

PROPERTIES OF MATTER 48

GASES AND VAPORS 50

USING GAS PRESSURE 52

---

gas laser
A type of **laser** using a gas instead of a solid or liquid. Electrons moving between the electrodes of the laser pass their energy on to the gas atoms. An energized atom emits a ray of light, which in turn hits and energizes another atom, causing it to emit a further ray of light. The rays of light bounce off mirrors at each end of the gas laser, causing a build-up of light. Eventually the beam of light becomes strong enough to pass through a half-silvered mirror at one end of the device, producing a laser beam.

gas laws
Physical laws of the behavior of gases. They include **Boyle's law** and **Charles' law**, which are concerned with the relationships between the pressure (p), temperature (T) and volume (V) of an ideal (hypothetical) gas. These two laws can be combined to give the general or universal gas law, which may be expressed as: pV/T = constant. Van der Waals' law includes corrections for the non-ideal behavior of real gases. **Graham's law** concerns gaseous diffusion.

gas turbine

An engine in which burning fuel supplies hot gas to spin a turbine. In a typical gas turbine, a multivaned compressor draws in and compresses air. The compressed air enters a combustion chamber at high pressure, and fuel is sprayed in and ignited. The hot gases produced escape through the blades of (typically) two turbines and spin them around. One of the turbines drives the compressor; the other provides the external power that can be harnessed. All jet engines are modified gas turbines. The first gas turbines were developed in the late 1930s. *See also* **internal combustion engine**.

Geiger counter

A device used for detecting ionizing radiation and/or measuring its intensity by counting the number of ionizing particles produced. It detects the momentary current that passes between electrodes in a suitable gas when a nuclear particle or a radiation pulse causes the ionization of that gas. The electrodes are connected to electronic devices that enable the number of particles passing to be counted. The increased frequency of measured particles indicates the intensity of radiation. *See* **radioactivity**.

generator

Any machine (for instance, an alternator or dynamo) that converts mechanical energy into electricity .

geostationary orbit

For an Earth satellite, the orbit (at a height of 35,900 kilometers) in which it takes 24 hours to complete one orbit.

Graham's law

One of the gas laws. It states that the speed at which a gas diffuses (as through a porous membrane) is proportional to the square root of its density.

gramophone

An instrument for reproducing sound, using a needle (pick-up) in contact with a spiral groove on an **audio disk**. A motor-driven turntable rotates the disk at constant speed and the head of the needle (pick-up) is made to vibrate due to tiny undulations in the grooves of the disk. The vibrations are then converted to electrical signals via a transducer. After amplification, the signals pass to one or two loudspeakers, which covert them to sound. The traditional name for the device, formerly used in the US and by its inventor Thomas Edison, was phonograph.

Grand Unified Theory

See **fundamental forces**.

GRAVITY

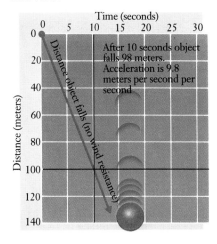

After 10 seconds object falls 98 meters. Acceleration is 9.8 meters per second per second

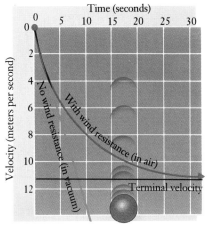

gravity

The natural force by which masses are drawn together by mutual attraction. Its mathematical definition was first given by Isaac Newton: "Any two particles of matter attract one another with a force directly proportional to the product of their masses and inversely proportional to the square of the distance between them." The force F of gravitational attraction between two objects of mass m_1 and m_2, separated by a distance d, is given as: $F = G\,(m_1 m_2/d^2)$, where G is the constant of gravitation and has the value $6.670 \times 10^{-11}\ \mathrm{N\ m^2\ kg^{-2}}$.

ground state

The lowest energy state of an atom or nucleus. It can be raised by **excitation**.

gyroscope

A mechanical istabilizing device, consisting of a heavy wheel mounted on an axis fixed in a ring that can be rotated around another axis, which is also fixed in a ring capable of rotation around a third axis. The components are arranged so that the axes of rotation in any position pass through the wheel's center of gravity. The wheel is thus capable of rotation around three mutually perpendicular axes, and its axis may take up any direction. If the axis of the spinning wheel is displaced, a restoring movement develops, returning it to its initial direction.

half-life

The time it takes for half of the **isotopes** in a given quantity of a radioactive substance to decay. Radioactive decay is exponential: the first 50 percent decays in the same time as the next 25 percent, and as the 12.5 percent after that, and so on. Carbon-14 takes about 5730 years for half the material to decay; another 5730 for half of the remaining half to decay; then 5730 years for half of that remaining half to decay; and so on. Other half-lives may be less than a second. In theory, decay is never complete and there is always some residual radioactivity. For this reason, the half-life is measured, rather than the total decay time. *See* **radioactivity**.

harmonic

An oscillation (such as an electromagnetic wave) whose frequency is a simple multiple of the fundamental sinusoidal oscillation (*see* **simple harmonic motion**). The fundamental frequency of a sinusoidal oscillation is the first harmonic. The second harmonic has a frequency of twice the fundamental.

heat

A form of internal energy possessed by a substance by virtue of the **kinetic energy** in the motion of its molecules or atoms. Heat is transferred by conduction, convection and radiation. It always flows from a region of higher temperature to one of lower temperature. Its effect on a substance may be simply to raise its temperature, or to cause it to expand, melt (if a solid), vaporize (if a liquid) or increase its pressure (if a confined gas).

heat capacity

The amount of heat that is needed to raise the temperature of an object by 1 degree. *See also* **specific heat capacity**.

heat engine

A device that converts heat into work. The heat is derived from the combustion of fuel. In an internal-combustion engine, the fuel is burnt inside the engine. In an external-combustion engine, such as a steam engine or steam turbine, the fuel is used to raise steam outside the engine and then some of the steam's internal energy is used to do work inside the engine. *See* **Carnot cycle**.

Heaviside layer

A layer of ionized gases in the atmosphere at an altitude of 110-120 kilometers, which reflects radio waves (and so allows long-range transmissions).

heavy water

See **nuclear reactor**.

hertz

The SI unit of frequency, equivalent to one cycle per second.

hologram

An image that contains three-dimensional information, produced by a photographic technique that splits a **laser** beam into two beams. Some holograms show meaningless patterns in ordinary light and produce a three-dimensional image only when laser light is projected through them. Reflection holograms produce images when ordinary light is reflected from them. Holographic techniques have applications in storing dental records, detecting strains in construction and in retail goods, and detecting forged paintings and documents.

Hooke's law

The law stating that the deformation of an object is proportional to the magnitude of the deforming force, provided that the object's elastic limit (*see* **elasticity**) is not exceeded. If the elastic limit is not reached, the object returns to its original size once the force is removed. For example, if a spring is stretched by 2 cm by a weight of 1 N, it will be stretched by 4 cm by a weight of 2 N, and so on; however, once the load exceeds the elastic limit of the metal in the spring, Hooke's law is no longer obeyed and each successive increase in weight results in a greater extension until the spring breaks.

humidity

The amount of water vapor in a gas (such as air), generally expressed as a percentage.

hydraulics

The study of the properties of water and other liquids, particularly the way they flow and transmit pressure, and the application of these properties in engineering. It applies the principles of hydrostatics and hydrodynamics. The oldest type of hydraulic machine is the hydraulic press, which consists of two liquid-connected pistons in cylinders, one of narrow bore, one of large bore. A force applied to the narrow piston applies a certain pressure (force per unit area) to the liquid, which is transmitted to the larger piston. Because the area of this piston is larger, the force exerted on it is larger. The original force has been magnified, although the smaller piston must move a great distance to move the larger piston only a little; mechanical efficiency is gained in force but lost in movement. The hydraulic principle of pressurized liquid increasing mechanical efficiency is commonly used on vehicle braking systems, bulldozers and the hydraulic systems of aircraft and excavators.

hydroelectricity

Electricity produced by generators driven by water turbines.

hydrometer

An instrument consisting of a weighted, long-stemmed glass bulb, which is floated in a liquid to measure its density.

hygrometer

An instrument for measuring humidity.

ideal gas

A hypothetical gas that, if it existed, would obey the gas laws exactly. An ideal gas would consist of molecules occupying very little space and would have insignificant forces between them. All collisions with the container or with other molecules would be perfectly elastic. *See* **gas laws**.

image

The representation of a physical object formed by a lens, a mirror or other optical instrument. Images may be of two kinds, real or virtual. A real image is formed by the convergence of rays that have passed through an image-forming device and can be projected onto a screen or photographic film. A virtual image is one from which rays appear to diverge and which therefore cannot be projected. Images may be upright or inverted, magnified or diminished.

CONNECTIONS

REFLECTION AND MIRRORS 110

REFRACTION AND LENSES 112

DISPERSION AND DIFFRACTION 114

TELEVISION CAMERA 124

TELEVISION RECEIVER 126

---

image orthicon

The most highly developed form of television camera tube. An image orthicon has an extreme sensitivity to light and can respond to light levels greatly below those that affect the human eye or that are used to expose cine film. Image orthicons are used in broadcasting because they give good image quality under most light levels encountered both outdoors and in television studios.

impedance

The total opposition of a circuit to the passage of an **electric current**. It has the symbol Z. For a **direct current (DC)** electric circuit it is simply the resistance (R). For **alternating current (AC)** it includes the reactance X (caused by capacitance or inductance); the impedance can be found using the equation $Z^2 = R^2 + X^2$, or $Z = R + iX$ where $i^2 = -1$. The term also refers to the ratio between the driving force and response of wave systems.

incandescence

The emission of light from a substance as a consequence of its high temperature. The **color** of the emitted **light** from liquids or solids depends on their temperature: in general, for solids, the higher the temperature the whiter the light. Gases may become incandescent through ionizing radiation, as in a glowing vacuum **discharge tube**.

inductance

A measure of the capability of an electric circuit or circuit component to form a magnetic field or store magnetic energy when carrying a current. Its symbol is L, and its SI unit is the henry.

induction

An alteration in the physical properties of an object brought about by the influence of a magnetic field. *See* **inductance**.

induction motor

An electric motor that employs an alternating current (AC). It comprises a stationary current-carrying coil (stator) surrounding another coil (rotor), which rotates because of the current induced in it by the magnetic field created by the stator; it does not require a commutator. *See* **electric motor**.

inertia

The tendency of an object to remain in a state of rest or uniform motion until an external force is applied, as stated by Isaac Newton's first law of motion (*see* **Newton's laws of motion**). The **mass** of an object is a measure of its inertia.

infrared radiation

Invisible electromagnetic radiation of wavelength between about 75 micrometers and 1 mm – that is, between the limit of the red end of the visible spectrum and the shortest microwaves. All objects above the absolute zero of temperature absorb and radiate infrared radiation. It is used in medical photography and treatment, and in industry, astronomy and criminology.

infrasound

Sound-like waves of frequencies below the usual audible limit of 20 Hz; "infra" means "below." *See also* **ultrasound**.

insulator

Any material that is a poor conductor of heat, sound or electricity. Most substances lacking free (mobile) electrons, such as non-metals, are electrical or thermal insulators. Glass or porcelain are particularly good insulators and are often used for insulating and supporting high-voltage electric cables. *See* **conductor**.

integrated circuit

A miniaturized electronic circuit produced on a single crystal, or chip, of a semiconducting material – usually silicon. It may contain many thousands of components and yet measure only 5 mm (0.2 in) square and 1 mm thick. Integrated circuits are usually enclosed in a plastic or ceramic case and linked via

INTERFERENCE

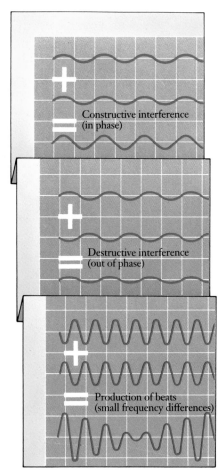

Constructive interference
(in phase)

Destructive interference
(out of phase)

Production of beats
(small frequency differences)

gold wires to metal pins, with which they are connected to a printed circuit board and the other components that make up such electronic devices as computers and calculators.

intensity

The power (or energy per second) per unit area carried by a form of radiation or wave motion. It is an indication of the concentration of energy present and can be measured at various distances from the source. For example, the intensity of light is a measure of its brightness, and may be shown to diminish with distance from its source in inverse proportion to the square of the distance.

interference

The phenomenon of two or more wave motions combining to produce a resultant wave of larger or smaller amplitude, depending on whether the combining waves are in or out of phase with each other. Interference may occur with light waves and sound waves.

CONNECTIONS

PRODUCING SOUND **98**

DISPERSION AND DIFFRACTION **114**

---

internal combustion engine

Any heat engine in which fuel is burned inside the engine, contrasting with an external combustion engine (such as a steam engine) in which fuel is burned in a separate unit. Diesel and gasoline engines are both internal combustion engines. They are reciprocating piston engines, in which pistons move up and down in cylinders to effect the engine operating cycle. This may be a **four-stroke cycle** or a two-stroke cycle. Gas turbines and rocket engines are sometimes also considered to be internal combustion engines because they burn their fuel inside their combustion chambers.

inverse-square law

The statement that the magnitude of an effect (usually a force) at a point is inversely proportional to the square of the distance between that point and the location of the force that causes it. Light, sound, gravitational force (Newton's law; *see* **gravity**), electrostatic force (**Coulomb's law**) and magnetic force (*see* **magnetism**) all obey the inverse-square law.

ion

An atom, or group of atoms, which is either positively charged (cation) or negatively charged (anion), as a result of the loss or gain of electrons during chemical reactions or exposure to certain forms of ionizing radiation.

ISOTOPE

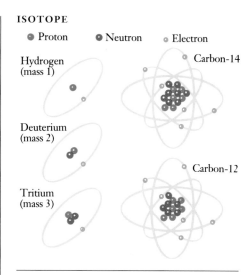

● Proton　　● Neutron　　○ Electron

Hydrogen (mass 1)

Deuterium (mass 2)

Tritium (mass 3)

Carbon-14

Carbon-12

ionizing radiation

Any short-wavelength electromagnetic radiation (for example, ultraviolet, X rays or gamma rays) or high-energy particles (electrons, protons or alpha particles) that produces ion pairs (ionization) when they pass through a medium.

ionosphere

The ionized layer of Earth's outer atmosphere (60-1000 km or 38-620 miles) that contains an appreciable concentration of ions and free electrons which modify the way in which radio waves are propagated, for instance by reflecting them back to Earth. The ionosphere is thought to be produced by absorption of the Sun's ultraviolet radiation, and shows daily and seasonal variations.

isothermal

Having the same temperature.

isotope

One of two or more elements that have the same atomic number (same number of protons in their atomic nuclei) but different numbers of neutrons, thus differing in their atomic masses. For example, hydrogen (one proton, no neutrons), deuterium (one proton, one neutron) and tritium (one proton, two neutrons) are all isotopes of hydrogen. Isotopes may be stable or radioactive, naturally occurring or synthesized. Most elements in nature consist of a mixture of isotopes. *See also* **radioactivity**.

CONNECTIONS

INSIDE THE ATOM **128**

SUBATOMIC PARTICLES **130**

THE UNSTABLE ATOM **132**

NUCLEAR FISSION **134**

NUCLEAR FUSION **136**

---

jet engine
See **gas turbine**.

joule
The SI unit of energy or work, equal to the work done when a force of one newton moves one meter toward the force.

Joule-Kelvin effect
The decrease in the temperature of a gas when subjected to **adiabatic expansion** into a vacuum ("fre expansion") through a porous plug or similar device. This due to the energy being used to overcome the cohesion of molecules of the gas. The effect was discovered by James Joule, working in collaboration with William Thompson (later Lord Kelvin), and is also known as the Joule-Thompson effect.

Kelvin scale
A temperature scale based on the thermodynamic principles of the performance of a reversible heat engine and the point at which the three phases of water (vapor, liquid and solid) coexist. The scale cannot have negative values and so it begins at **absolute zero** (-273.15°C) and increases by the same degree intervals as the Celsius scale. Its unit is the kelvin (the SI unit of temperature).

kinetic energy
The energy of an object resulting from motion. It is contrasted with **potential energy**.

kinetic theory
The theory describing the physical properties of matter in terms of the behavior (such as movement) of its component atoms or molecules. A gas consists of rapidly moving atoms or molecules. According to kinetic theory, it is their continual impact on the walls of the containing vessel that accounts for the pressure of the gas (see **Brownian motion**); and the slowing of molecular motion as temperature falls accounts for the physical properties of liquids and solids, culminating in no molecular motion at **absolute zero**. By making various assumptions about the nature of gas molecules, it is possible to derive the various gas laws from kinetic theory. See **gas laws**.

Kirchhoff's laws
In a complex electric circuit, **1** the sum of the currents flowing to any junction is zero, and **2** around any closed circuit, the sum of the electromotive forces (e.m.f.s) equals the sum of the products of individual current and impedances.

laser
An acronym for **l**ight **a**mplification by **st**imulated **e**mission of **r**adiation. A laser is a device for producing a narrow beam of light of the same wavelength and phase (coherent light), capable of traveling over vast distances without dispersion and of being focused to give enormous power densities. Any material in which the majority of atoms or molecules can be excited can be used as to make a laser. Early lasers used crystals which gave a high-power, pulsed output. Helium-neon gas lasers are capable of producing a continuous output, but at a lower power. Lasers have many applications in industry and medicine. See also **hologram** and **maser.**

latent heat
The heat required to change the physical state of a substance – such as from solid to liquid (fusion) or from liquid to gas (vaporization) without changing its temperature. Most substances have a latent heat of fusion and a latent heat of vaporization.

lattice
1 A regularly spaced arrangement of atoms, ions or molecules in a crystalline solid. **2** The regular discrete geometrical pattern of fissionable and non-fissionable material in a nuclear reactor.

Leclanché cell
A primary cell that has a zinc cathode and carbon anode dipping into an electrolyte of ammonium chloride solution. It is the basis of the common form of dry cell (battery).

left-hand rule
If the directions of the thumb, first finger and second finger of the left hand are held at right angles, the thumb indicates the direction of movement, the first finger the direction of the magnetic field and the second finger the direction of induced current flow in an electric motor.

lens
A piece of a transparent material, such as glass, with two polished surfaces – one concave or convex, and the other plane, concave, or convex – that modifies rays of light. A convex (positive) lens brings rays of light together; a concave (negative) lens makes the rays diverge. Lenses are essential to spectacles, microscopes, telescopes, cameras and all optical instruments.

Lenz's law
For a wire moving in a magnetic field, the electric current thereby induced in the wire itself generates a magnetic field that tends to oppose the movement.

lever
A simple machine that consists of a rigid rod pivoted at a fixed point called the fulcrum, used for shifting or raising a heavy load or applying force in a similar way. Levers are classified according to where the effort is applied and the load moving force developed in relation to the position of the fulcrum.

light
The form of electromagnetic radiation to which the human eye is sensitive, having a wavelength from about 400 nm in the extreme violet to about 700 nm in the extreme red. Light is considered to exhibit both particle and wave properties, and the fundamental particle, or quantum, of light is called the photon. The speed of light (and all electromagnetic radiation) in a vacuum is 299,792.5 km/sec (186,281 miles/sec) and is a universal constant.

CONNECTIONS

LIGHT AND THE SPECTRUM **106**

PRODUCING LIGHT **108**

DISPERSION AND DIFFRACTION **114**

LIGHT FROM LASERS **116**

---

linear motor
An electric induction motor in which the fixed stator and moving armature are straight and parallel to each other (rather than being circular and one inside the other as in an ordinary induction motor). A magnetic force exists between the stator and the armature. Linear motors are used to power high-speed trains that float above the track. See **maglev**.

line of force
An imaginary line drawn in an electric or magnetic field so that its direction at every point gives the direction of the electric or magnetic force at that point.

KINETIC THEORY

Gas

Solid

Liquid

LOUDSPEAKER

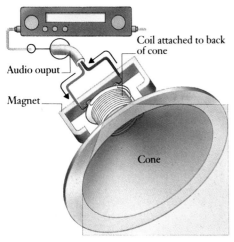

Coil attached to back of cone

Audio ouput

Magnet

Cone

liquefaction of gases
A gas must be cooled below its critical temperature to be liquefied by pressure alone.

liquid
A state of matter between a solid and a gas, in which the shape of a given mass depends on the vessel that contains it, the volume being independent. A liquid is practically incompressible – a property made use of in hydraulics.

liquid crystal display
The display of numbers (for example, in a calculator) or pictures (such as on a pocket television screen) produced by molecules of a substance in a semiliquid state with some crystalline properties, so that clusters of molecules form orderly alignments. The liquid crystals are sandwiched between transparent sheets of conductive material. The display is a blank until the application of an electric field, which "twists" the molecules so that they reflect or transmit light falling on them. Little power is consumed and edge-lighting can be provided for use in darkness.

load
The output force of any machine. *See* **distance ratio** and **force ratio**.

Lorenz-Fitzgerald contraction
When an object moves at near the speed of light, its length decreases in the direction of motion, relative to the frame of reference from which the measurement is made. This discovery helped lay the foundations for the special theory of **relativity**.

loudness
The intensity or volume of a sound as perceived by the human ear. Because the ear responds differently to different frequencies, loudness is dependent on frequency. Sounds with frequencies between 1000 and 5000 Hz are louder than sounds at the same intensity at higher or lower frequencies.

loudspeaker
An electromechanical device (transducer) that converts electrical signals into sound waves. In a moving-coil loudspeaker, electrical signals are fed to a coil of fine wire wound around the top of a cone. The coil is surrounded by a magnet. When signals pass through it, the coil becomes an electromagnet, which by moving causes the cone to vibrate, setting up sound waves.

luminescence
The emission of light from an object when its atoms are excited by means other than raising its temperature. Short-lived luminescence is called **fluorescence**; longer-lived luminescence is called **phosphorescence**. When exposed to an external source of energy, the outer electrons in atoms of a luminescent substance absorb energy (*see* **excitation**) and "jump" to a higher energy level. When these electrons return to their former level they emit their excess energy as light. *See also* **bioluminescence**.

machine
Any device that allows a small force (the effort) to overcome a larger one (the load). There are three basic machines: the inclined plane (ramp), the lever, and the wheel and axle. All other machines are combinations of these three basic types. Simple machines derived from the inclined plane include the wedge, the gear and the screw; the spanner is derived from the lever; the pulley from the wheel and axle. The principal features of a machine are its **force ratio**, which is the ratio of load to effort; its **distance ratio**; and its **efficiency**, which is the work done by the load divided by the work done by the effort. Efficiency is expressed as a percentage. In a perfect machine, with no friction, the efficiency would be 100 percent. All practical machines have efficiencies of less than 100 percent, otherwise perpetual motion would be possible.

CONNECTIONS

MECHANICAL ENERGY **64**

SIMPLE MACHINES **66**

MEASURING AND USING HEAT **72**

ELECTRIC MOTORS **86**

Mach number
The ratio of the speed of an object to the speed of sound in the undisturbed medium through which the object travels, under the same conditions of temperature and pressure. Mach 1 is reached when an object (such as an aircraft) has a velocity greater than that of sound (passes the sound barrier) – that is, 331 m/sec (1087 ft/sec) at sea level. The Mach number is named for the Austrian physicist Ernst Mach.

maglev
An acronym derived from **mag**netic **lev**itation. Maglev is the name given to a type of high-speed surface transport using the repellent force of superconductive magnets (*see* **superconductivity**) to propel and support. A high-speed train above a track is levitated by electromagnets and forward thrust is provided by linear motors aboard the cars, propelling the train along a reaction plate.

magnet
Any object that forms a magnetic field (displays **magnetism**), either permanently or temporarily through induction, causing it to attract materials such as iron, cobalt, nickel and alloys of these. A magnet always has two magnetic poles, called north and south.

magnetic field
The region around a permanent magnet, or around a conductor carrying an electric current, in which a force acts on a moving charge or on a magnet placed in the field. The field can be represented by lines of force, which by convention link north and south poles and are parallel to the directions of a compass needle placed on them. Its magnitude and direction are given by the magnetic flux density, expressed in teslas. Magnetic fields originate at magnetic dipoles or electric currents.

magnetic induction
Magnetization induced in a magnetic material (by placing it in a magnetic field).

MAGNETIC FIELD

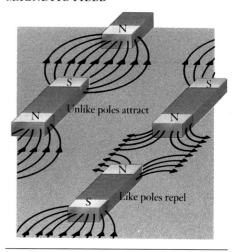

Unlike poles attract

Like poles repel

magnetic moment

A measure of the magnetic strength of an atom, molecule, moving charge, permanent magnet or a coil carrying an electric current. The torque (turning force) on the magnet equals the product of the magnetic moment and the magnetic induction.

magnetic north

The direction in which the north-seeking pole of a freely pivoted magnet (such as a compass needle) points. It differs from the geographical north by an angle called the magnetic declination.

magnetic pole

The region on a magnet where its magnetic effects are strongest. Magnets always have two poles, called north and south. When a magnet is suspended freely, the north-seeking pole always points to magnetic north and the south pole always points south. The north pole of one magnet is attracted to the south pole of another but repelled by its north pole. Like poles attract, unlike repel.

magnetic tape

Flexible plastic tape, typically 6 to 50 mm wide, coated with a magnetizable material (usually ferromagnetic iron oxide) on which electrical signals are recorded for subsequent reproduction. It is used to store audiovisual signals, sound and computer data.

magnetism

The phenomenon associated with magnetic fields. Magnetic fields are produced by moving charged particles: in electromagnets, electrons flow through a coil of wire connected to a battery; in permanent magnets, spinning electrons within the atoms generate the field. Substances differ in the extent to which they can be magnetized by an external field (susceptibility). Materials that can be strongly magnetized, such as iron, cobalt and nickel, are said to be ferromagnetic; this is due to the alignment of molecular magnets in areas called domains. Ferromagnetic materials lose their magnetism if heated to the critical temperature. Most other materials are paramagnetic, being only weakly pulled toward a strong magnet. This is because their atoms have a low level of magnetism and do not form permanent domains. Diamagnetic materials are weakly repelled by a magnet since electrons within their atoms act as electromagnets and oppose the applied magnetic force. Antiferromagnetic materials have a very low susceptibility that increases with temperature; a similar phenomenon in materials such as ferrites is called ferrimagnetism. Apart from its universal application in dynamos, electric motors and switch gears, magnetism is of considerable importance in advanced technology, such as in particle accelerators for nuclear research, memory stores for computers, tape recorders and cryogenics.

CONNECTIONS

ELECTRICITY AND MAGNETISM 74

MAGNETS AND FIELDS 76

ELECTRIC MOTORS 86

---

magnetohydrodynamics (MHD)

The field of science concerned with the behavior of ionized gases or liquid in a magnetic field. The movement of the gases or liquids gives rise to induced electric currents which interact with the magnetic field, modifying the motion in turn. The phenomenon can be used to generate electrical power.

magnification

The measure of the enlargement or reduction of an object in an imaging optical system. Linear magnification is the ratio of the size (height) of the image to that of the object. Angular magnification is the ratio of the angle subtended at the observer's eye by the image to the angle subtended by the object when viewed directly.

maser

An acronym for **m**icrowave **a**mplification by **s**timulated **e**mission of **r**adiation. A maser is a high-frequency microwave amplifier or oscillator in which the signal to be amplified is used to stimulate unstable atoms into emitting energy at the same frequency. Atoms or molecules are raised to a higher energy level (*see* **excitation**) and then allowed to lose this energy by radiation emitted at a precise frequency. *See also* **laser**.

mass

The amount of substance that an object contains. The force of **gravity** acting upon mass produces weight. Thus, the weight of an object on the Moon will be less than on Earth, while its mass remains the same. Mass is a scalar quantity, meaning that it is represented by a real number and has no direction (it is not a vector).

mass defect

The difference between the mass of an atomic nucleus and the total mass of all the particles that make it up.

mass number

The numbers of protons and neutrons in the nucleus of an atom, each taken as a unit of mass. It is used along with the atomic number (the number of protons) in nuclear notation: in symbols that represent isotopes, such as $^{14}_{6}C$, the lower number is the atomic number and the upper is the mass number.

MASS SPECTROMETER

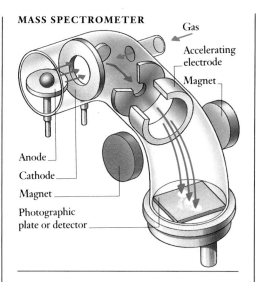

Gas

Accelerating electrode

Magnet

Anode

Cathode

Magnet

Photographic plate or detector

mass spectrometer

An analytical device for determining the chemical composition of a substance. Positive ions (charged particles) are separated by an electromagnetic system, which permits accurate measurement of the relative concentrations of the various ionic masses present, particularly isotopes. It consists essentially of a source of positive ions (often produced by bombarding gaseous atoms with electrons) which are accelerated by an electric field and deflected by a magnetic field to varying extents depending on their charge and mass.

matter

Anything that has mass and can be detected and measured. All matter is made up of atoms, which in turn are made up of elementary particles; it exists ordinarily as a solid, liquid or gas. The history of science and philosophy is largely concerned with accounts of theories of matter, ranging from the hard "atoms" of Democritus to the "waves" of modern quantum theory.

mechanical advantage

See **force ratio**.

mechanical equivalent of heat

A constant factor relating the calorie (the c.g.s. unit of heat) to the joule (the unit of mechanical energy), equal to 4.186 joules per calorie. It is unnecessary in the SI system of units, which measures heat and all forms of energy in joules (so that the mechanical equivalent of heat is 1). It is now simply recognized as the specific heat capacity of water, $4.186 \text{ kJ kg}^{-1}\text{K}^{-1}$.

mechanics

The branch of physics that deals with forces and matter, consisting of statics (the effects of forces that do not produce motion), dynamics (effects of forces on moving objects) and kinematics (velocity and acceleration).

melting point

The temperature at which a substance melts, or changes from a solid to liquid form. A pure substance under standard conditions of pressure (usually one atmosphere) has a definite melting point. If heat is supplied to a solid at its melting point, the temperature does not change until melting is complete. The heat supplied is the latent heat of fusion.

meniscus

The curved shape of the surface of a liquid in a thin tube, caused by the cohesive effects of surface tension (capillary action). When the walls of the container are made wet by the liquid, the meniscus is concave, but with highly viscous liquids (such as mercury) the meniscus is convex. Meniscus also refers to a concavo-convex or convexo-concave lens.

mirror

Device that reflects light. Mirrors are commonly plane (flat), concave (domed inward) or convex (domed outward).

microminiaturization

The gradual decrease in the size of electronic circuit components, from the early thermionic tube (valve) to the transistor, to the integrated circuit and the large-scale integrated (LSI) circuit, to the very large-scale integrated (VLSI) circuits of the present day.

microphone

A device that converts sound waves into electrical signals using a transducer. Many telephones have a carbon microphone, which reproduces a narrow range of frequencies. For live music, a moving-coil microphone is often used, with a diaphragm that vibrates with sound waves moves a coil in a magnetic field to generate an electric current. A crystal microphone makes use of the **piezoelectric effect.**

microprocessor

In a microcomputer, a silicon chip that contains the arithmetic unit and logic function of the central processing unit (CPU).

microscope

An instrument for magnification with high resolution for detail. Optical and electron microscopes are the ones most widely used. An optical microscope usually has two sets of glass lenses and an eyepiece. Fluorescence microscopy makes use of fluorescent dyes to illuminate samples, or to highlight the presence of particular substances within a sample. A transmission electron microscope passes a beam of electrons, instead of a beam of light, through a specimen. Because electrons are not visible, the eyepiece is replaced by a fluorescent screen or photographic plate; far higher magnification and resolution are possible than with the optical microscope. A scanning electron microscope moves a fine beam of electrons over the surface of a specimen, collecting the reflected electrons to form the image. The specimen has to be in a vacuum chamber. An acoustic microscope passes ultrasonic waves through the specimen, the transmitted sound being used to form an image on a computer screen.

A scanned-probe microscope runs a probe, with a tip that may consist of a single atom, across the surface of the specimen, which requires no special preparation. In a scanning tunneling microscope, an electric current that flows through the probe is used to construct an image of the specimen. In an atomic force microscope, the force felt by the probe is measured and forms the image. These instruments can magnify a million times and give images of single atoms.

microwaves

Electromagnetic waves with wavelengths in the range 0.3-30 cm or frequencies of 300-300,000 megahertz (between radio waves and infrared radiation). They are used in **radar**, as carrier waves in radio broadcasting, and in microwave heating and cooking.

moderator

A material such as graphite or heavy water used in a **nuclear reactor** to reduce the speed of high-energy neutrons. Neutrons produced by nuclear fission are fast-moving and must be slowed to initiate further fission and sustain a **chain reaction** so that nuclear energy is released at a controlled rate.

modulation

The process of superimposing an intermittent change of frequency, or amplitude, on a radio carrier wave, in accordance with the audio characteristics of the speaking voice, music or other signal being transmitted. Laser light can also be modulated to carry information. *See* **amplitude modulation (AM)** and **frequency modulation (FM).**

molecule

A group of two or more atoms bonded together. A molecule of an element consists of similar atoms; a molecule of a compound consists of two or more different atoms. Molecules vary in size and complexity from the hydrogen molecule (H_2) to the macromolecules of proteins and plastics. They are held together by ionic bonds, in which the atoms gain or lose electrons to form **ions**, or by covalent bonds, in which electrons from each atom are shared in a new **orbital**.

moment

The measure of the turning effect (torque) produced by force acting on an object. It is equal to the product of the force and the perpendicular distance from its line of action to the point around which the object turns. The unit is the newton meter. *See* **couple**.

momentum

The product of the mass of an object and its velocity. The momentum of an object does not change unless it is acted on by an external force. The law of conservation of momentum is one of the fundamental concepts of classical physics. It states that the total momentum of all objects in a closed system is constant and remains unaffected by processes occurring within the system. *See also* **angular momentum**.

monochromatic light

Light containing radiation of a single wavelength. No source emits truly monochromatic light, although very narrow frequency bands of wavelengths can be obtained, such as with discharge tubes (mercury, neon and sodium lighting) and lasers.

MODULATION

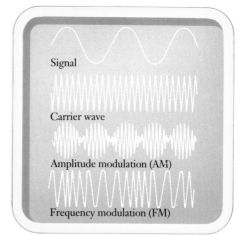

Signal

Carrier wave

Amplitude modulation (AM)

Frequency modulation (FM)

neutron

One of the three chief subatomic particles (the others being the proton and the electron). Neutrons have about the same mass as protons but no electric charge, and occur in the nuclei of all atoms except hydrogen. They contribute to the mass of atoms but do not affect their chemistry, which depends on the proton or electron numbers. Isotopes of a single element differ only in the number of neutrons in their nuclei and have identical chemical properties. Outside a nucleus, a neutron is unstable, decaying to give a proton and an electron.

> **CONNECTIONS**
>
> INSIDE THE ATOM **128**
> NUCLEAR FISSION **134**

neutron number

The number of neutrons in the nucleus of a particular atom, equal to the **nucleon number** minus the **atomic number**. It is significant in determining the stability of the nucleus; nuclei with an excess of neutrons tend to be radioactive.

Newton's laws of motion

The three laws that form the basis of Newtonian (classical) mechanics. **1** Unless acted upon by a net force, an object at rest stays at rest and a moving object continues moving at the same speed in the same straight line. **2** A net force applied to an object gives it a rate of change of momentum proportional to the force and in the direction of the force. **3** When an object A exerts a force on an object B, B exerts an equal and opposite force on A – that is, for every action there is an equal and opposite reaction. These laws were first expressed by the English mathematician and philosopher Isaac Newton in his *Principia* of 1687.

Newton's rings

An optical **interference** phenomenon, seen as concentric rings of spectral colors where white light passes through a thin film of transparent medium, such as the wedge of air between a large convex lens and a flat glass plate. With monochromatic light the rings appear as alternate light and dark bands. They are caused by interference between light rays reflected from the plate and those reflected from the curved surface of the lens.

normal

A line (or plane) that is at right angles to another line (or plane). *See also* **angle of incidence**.

n-type semiconductor

A type of semiconductor that carries electric current by means of moving electrons.

nuclear fission

The process in which an atomic nucleus breaks up into two or more major fragments with the emission of neutrons. It is accompanied by the release of energy in the form of gamma radiation and the kinetic energy of the emitted particles, manifested as heat. Fission occurs spontaneously in some nuclei, or can be induced by bombarding nuclei (such as those of uranium-235) with neutrons. The neutrons released spontaneously by the fission of the uranium nuclei may be used in turn to induce further fissions, setting up a **chain reaction**.

> **CONNECTIONS**
>
> INSIDE THE ATOM **128**
> NUCLEAR FISSION **134**
> MASS AND ENERGY **142**

nuclear fusion

The process in which two atomic nuclei are fused, with the release of a large amount of energy. Very high temperatures and pressures are required in order for the process to happen. Under these conditions the atoms involved are stripped of all their electrons so that the remaining particles, which together make up plasma, can come close together at very high speeds and overcome the mutual repulsion of the positive charges on the atomic nuclei. At very close range another nuclear force comes into play, fusing the particles together to form a larger nucleus.

nuclear reactor

A device for producing nuclear energy in a controlled manner. There are various types of reactor, all of which use nuclear fission. In a gas-cooled reactor, a circulating gas under pressure removes heat from the core of the reactor, which usually contains natural uranium. The efficiency of the fission process is increased by slowing neutrons in the core by using a **moderator**. The reaction is controlled with neutron-absorbing rods made of boron. An advanced gas-cooled reactor generally has enriched uranium as its fuel. A water-cooled reactor, such as the steam-generating heavy water (deuterium oxide) reactor, has water circulating through the hot core. The water is converted to steam, which drives turbo-alternators to generate electricity. The most widely used reactor is the pressurized-water reactor, which contains a sealed system of pressurized water that is heated to form steam in heat exchang-

ers in an external circuit. A fast reactor has no moderator and uses fast neutrons to bring about fission. It uses a mixture of plutonium and uranium oxide as fuel. When the reactor is operating, uranium is converted to plutonium, which can be extracted and used later as fuel. The fast breeder produces more plutonium than it consumes. Heat is removed by a coolant of liquid sodium.

nucleon number

The number of nucleons (neutrons or protons) in the nucleus of a particular atom, also known as the **mass number**.

nucleus

The positively charged central part of an atom which constitutes almost all its mass. Except for a hydrogen nucleus, which consists of a single proton, nuclei are composed of both protons and neutrons. Nuclei are surrounded by electrons, which carry a negative charge equal to the protons, thus giving the atom a neutral charge.

Ohm's law

The law that states that the current flowing in a metallic conductor maintained at constant temperature is directly proportional to the potential difference (voltage) between its ends. If a current of I amperes flows between two points in a conductor across which the potential difference is V volts, then V/I is a constant called the resistance R ohms between those two points. As an equation, it is written as $V/I = R$ or $V = IR$.

optics

A branch of physics that deals with the study of light and vision – for example, shadows and mirror images, lenses, microscopes, telescopes and cameras. Light rays are thought of as traveling in straight lines, although Albert Einstein demonstrated that they may be "bent" by a gravitational field. On striking a surface they are reflected or refracted with some absorption of energy, and the study of this is known as geometrical optics.

> **CONNECTIONS**
>
> LIGHT AND THE SPECTRUM **106**
> REFLECTION AND MIRRORS **110**
> REFRACTION AND LENSES **112**
> DISPERSION AND DIFFRACTION **114**
> TELEVISION CAMERA **124**

orbit

1 The path of an electron as it travels around the nucleus of an atom (*see* **orbital**). **2** The path of a planet around a star, or of a moon or satellite around a planet.

ORBITAL

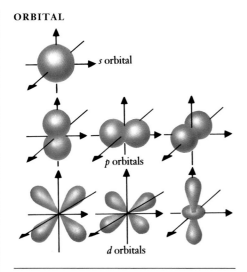

s orbital

p orbitals

d orbitals

orbital

The region around the nucleus of an atom (or, in a molecule, around several nuclei) in which an electron is likely to be located. According to quantum theory, the position of an electron is uncertain; it may be located at any point. However, it is more likely to be found in some places than in others and it is these that make up the orbital. An atom or molecule has numerous orbitals, each of which has a fixed size and shape. For example, the innermost orbital – the *s* orbital – is spherical. There are up to three *p* orbitals, which have a dumbbell shape. Four of the five possible *d* orbitals resemble swollen crosses; the fifth is an elongated dumbbell surrounded by a central "doughnut". Three quantum numbers, representing energy, angular momentum and orientation, uniquely characterize each type of orbital.

oscillator

Any device that produces a desired oscillation (vibration). There are many types of oscillator for different purposes. An oscillator is an essential part of a radio transmitter, generating the high-frequency carrier signal necessary for radio communication. The frequency is often controlled by the vibrations set up in a crystal (such as quartz).

osmosis

The movement of a solvent from a dilute solution to a more concentrated solution across a semipermeable membrane.

parallel circuit

An electrical circuit in which current is split between two or more parallel paths. The division of the current across each conductor is in the ratio of their resistances. If the currents across two conductors of resistance R_1 and R_2, connected in parallel, are I_1 and I_2, then the ratio of those currents is given

by the equation: $I_1/I_2 = R_2/R_1$. The total resistance R of those conductors is expressed as: $1/R = 1/R_1 + 1/R_2$. *See* **series circuit**.

paramagnetism

The small positive susceptibility of a substance, stronger than diamagnetism but weaker than ferromagnetism. Its atoms or molecules have net orbital spin or spin magnetic moments that are capable of being aligned in the direction of an applied magnetic field. Paramagnetism occurs in all atoms and molecules with unpaired electrons, and in metals; it varies inversely with temperature. *See* **magnetism**.

particle accelerator

See **accelerator**.

Pascal's law

Pressure applied at any point in a fluid is transmitted equally throughout the fluid.

Pauli exclusion principle

In any atom, no two electrons can have the same set of **quantum numbers**, so there can be a maximum of two electrons in any one orbital.

pendulum

A weight at the end of a string which swings (through small angles) with simple harmonic motion with a period equal to 2π times the square root of its length divided by the **acceleration** of free fall.

perfect gas

See **ideal gas**.

permanent magnet

A magnet made of ferromagnetic material, which retains its magnetization although the magnetizing field has been removed. *See* **ferromagnetism** and **magnetism**.

PHOTOCELL

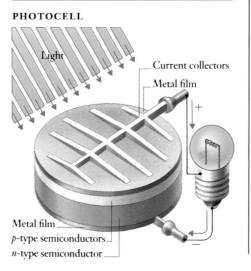

Light

Current collectors

Metal film

Metal film
p-type semiconductors
n-type semiconductor

phase

A stage in an oscillating motion, such as a wave motion: two waves are in phase when their peaks and their troughs coincide. Otherwise, they are out of phase; the phase difference has consequences in interference phenomena and alternating current electricity. Any physically distinct (and homogeneous) part of a chemical system, separated from the rest of the system by definite boundaries, is also called a phase. Water and ice are two phases of water that can coexist.

phosphor

Any substance that is phosphorescent – that is, gives out visible light when it is illuminated by a beam of electrons or ultraviolet light. Television screens (*see* **cathode-ray tube**) are coated on the inside with phosphors that glow when beams of electrons strike them. Fluorescent lamp tubes also have an internal phosphor coating. Phosphors are also used in Day-Glo paints and as optical brighteners in detergents.

phosphorescence

The emission of light by certain substances after they have absorbed energy, whether from visible light, other electromagnetic radiation (such as ultraviolet rays or X rays), or cathode rays (a beam of electrons). When the stimulating energy is removed, phosphorescence ceases, although it may persist for a short time (unlike fluorescence, which stops immediately).

photocell

A device that produces an electric current when exposed to light. In a photoemissive cell, the radiation causes electrons to be emitted and a current to flow (**photoelectric effect**); a photovoltaic cell causes an electromotive force to be generated in the presence of light across the boundary of two substances. A photoconductive cell, which contains a semiconductor, increases its conductivity when exposed to electromagnetic radiation.

photoelectric effect

The emission of electrons from a substance (usually a metallic surface) when it is struck by photons, either those of visible light or ultraviolet radiation. The energy of the emitted electrons depends on the frequency of the incident radiation: the higher the frequency, the greater the energy of its photons. The number of electrons emitted depends on the radiation's intensity.

photoelectricity

Electricity produced by utilizing the photoelectric effect.

photon

The elementary particle or "package" (quantum) of energy in which light and other forms of electromagnetic radiation are emitted. The photon has both particle and wave properties; it has no charge and is considered massless, but it possesses momentum and energy. The photon cannot be subdivided and is the carrier of the electromagnetic force, one of the fundamental forces of nature. According to **quantum theory**, the **energy** of a photon is given by the formula $E = h\nu$, where h is Planck's constant and ν is the frequency of the radiation.

photoresist process

The process of selectively removing the oxidized surface of a silicon chip semiconductor. The photoresist material is usually an organic substance that polymerizes on exposure to ultraviolet light; in that form it is resistant to attack by acids and solvents.

pick-up

A mechanical-electrical device actuated by a diamond or graphite stylus which rests on the insides of the grooves on an audio disk. The pick-up (needle) drives an audio amplifier by generating a voltage corresponding to the tracking of the groove (or, for stereo reproduction, two voltages). *See* **transducer**.

picture tube

A **cathode-ray tube** used for displaying television images or other waveform data for the purpose of information or control.

piezoelectric effect

Electricity produced across the opposite faces of some crystals (such as quartz) when they are subjected to a mechanical strain. Conversely, the crystals expand or contract in size when they are subjected to an external voltage. Piezoelectric crystal oscillators are used as frequency standards (such as in watches) and in phonograph needles and some microphones.

piston

A barrel-shaped device used in reciprocating engines (steam, gasoline, diesel) to harness power. Pistons are driven up and down in cylinders by expanding steam or hot gases. They pass on their motion via a connecting rod and crank to a crankshaft, which rotates to drive machinery. In a pump or compressor, the role of the piston is reversed; it is used instead to move gases and liquids. *See also* **internal-combustion engine**.

pitch

The property of sound that characterizes its highness or lowness. Pitch is related to frequency, but they are not identical. In standard pitch, A above middle C has a frequency of 440 Hz. The pitch of a note is also related to its loudness.

pixel

An acronym for picture element. A pixel is the smallest controllable element on a video screen display. All screen images are made up of a collection of pixels, with each pixel being either off (dark) or on (illuminated, possibly in color). The number of pixels determines the screen's resolution.

Planck's constant

A fundamental physical constant equal to 6.6256×10^{-34} joule second. For a vibrating system, it is the ratio of its energy to its frequency. *See* **Planck's law of radiation**.

Planck's law of radiation

Radiant energy is absorbed or emitted by an object in multiples of a basic amount, called a quantum (that is, the energy is not absorbed or emitted continuously). *See also* **quantum theory**.

plasma

An ionized gas produced at extremely high temperatures, as in the Sun and other stars, which contains positive and negative charges in approximately equal numbers. It is a good electrical conductor. In fusion reactions the plasma produced is confined through the use of magnetic fields. *See* **nuclear fusion**.

polarized light

Light in which electromagnetic vibrations take place in one direction. In ordinary (unpolarized) light, the electric and magnetic fields vibrate in all directions perpendicular to the direction of propagation. After reflection from a polished surface or transmission through certain materials (such as Polaroid),

POLARIZED LIGHT

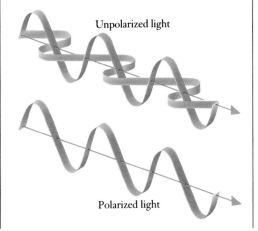

Unpolarized light

Polarized light

the electric and magnetic fields are confined to one direction and the light is said to be plane polarized. In circularly polarized light, the magnetic and electric fields are confined to one direction, but the direction rotates as the light propagates.

pole (magnetic)

See **magnetic pole**.

potential difference

The measure of the electrical potential energy converted to another form for every unit charge moving between two points in an electric circuit. The unit of potential difference is the volt. Potential difference V may be defined by: $V = W/Q$, where W is the electrical energy converted in joules and Q is the unit charge in coulombs. *See also* **electric potential** and **Ohm's law**.

potential energy

The energy of an object due to its position or state, without any observable change, such as an object raised against the pull of **gravity**. An object of mass m at a height h above the ground possesses potential energy mgh, where g is the acceleration due to gravity (acceleration of free fall), since this is the amount of work done in falling to the ground. An object in a state of tension or compression (such as a stretched spring) also has potential energy. *See* **kinetic energy**.

CONNECTIONS

FORCE AND ENERGY 58

FORCE OF GRAVITY 62

MECHANICAL ENERGY 64

SIMPLE MACHINES 66

---

power

The rate of doing work or consuming energy, measured in watts (joules per second) or other units of work per unit time. If the work done is W joules and the time taken is t seconds, the power P is given as: $P = W/t$.

pressure

The force acting normally (at right angles) to an object per unit surface area. The SI unit of pressure is the pascal (Pa or newton per square meter), equal to 0.01 millibars. In a liquid or gas, pressure increases with depth. Pressure at a depth h in a fluid of density d is equal to hdg, where g is the acceleration due to gravity (aceleration of free fall). At the edge of Earth's atmosphere, atmospheric pressure is zero, whereas at sea level atmospheric pressure is about 100 kPa (1013 millibars or 1 atmosphere). Atmospheric pressure is commonly measured by a barometer.

primary cell

An electrical cell that cannot be replenished and so, after prolonged use, becomes discharged or "dead." *See* **cell, battery, dry battery** and **secondary cell**.

primary color

For light, red, green and violet, which mix to give white and can be combined to give all other colors. For pigments, red, yellow and blue, which mix to give black and can be combined to make all other colors.

printed circuit

An electrical circuit created by printing "tracks" of a conductor such as copper on one or both sides of an insulating board. Components such as integrated circuits (chips), resistors and capacitors can be soldered to the surface of the board (surface-mounted) or, more commonly, attached by inserting their connecting pins or wires into holes drilled in the board and then soldering.

proton

A positively charged subatomic particle, a fundamental constituent of any atomic nucleus. Its lifespan is effectively infinite. A proton carries a unit positive charge equal to the negative charge of an electron. Its mass is almost 1,836 times that of an electron, or 1.67×10^{-24} grams. The number of protons in the atom of an element is equal to its **atomic number**.

p-type semiconductor

A type of semiconductor that carries electric current by means of moving positive holes.

quantum

General term for the indivisible unit of any form of energy. The word refers in particular to the photon, the individual quantity of electromagnetic radiation energy.

quantum number

One of a set of four numbers that uniquely characterize an electron and its state in an atom. The principal quantum number n defines the electron's main energy level. The orbital quantum number l relates to its angular momentum. The magnetic quantum number m describes the energies of electrons in a magnetic field. The spin quantum number m_s gives the spin direction of the electron. No two electrons in an atom can have the same set of quantum numbers – this is the Pauli exclusion principle.

quantum theory

The theory that many quantities, such as energy, cannot have a continuous range of values, but only a number of particular ones,

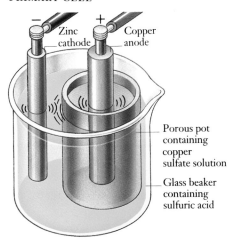

PRIMARY CELL

Zinc cathode — Copper anode
Porous pot containing copper sulfate solution
Glass beaker containing sulfuric acid

because they are packaged in "quanta of energy". Just as earlier theories showed how light, generally seen as a wave motion, could also in some ways be seen as composed of discrete particles (photons), quantum mechanics shows how atomic particles such as electrons may also be seen as having wave-like properties. Quantum mechanics is the basis of particle physics, modern theoretical chemistry and the solid-state physics that describes the behavior of the silicon chips used in computers. The theory began around 1900 with the work of Max Planck on radiated energy and was extended by Albert Einstein to electromagnetic radiation generally, including light. Niels Bohr used it to explain the spectrum of light emitted by excited hydrogen atoms. Later work by Schrödinger, Heisenberg, Dirac and others elaborated the theory to what is now called quantum mechanics (or **wave mechanics**).

CONNECTIONS

QUANTUM PHYSICS **138**

WAVES AND PARTICLES **140**

MASS AND ENERGY **142**

---

quark

The elementary subatomic particle that is the fundamental component of all neutrons and protons, but not electrons. There are six types or "flavors" of quark: up, down, top, bottom, strange and charm, each of which has three varieties or "colors": red, yellow and blue. (The quark is not actually colored, although the analogy is useful in many ways). For each quark there is an antiparticle, called an antiquark.

radar

An acronym for **ra**dio **d**irection **a**nd **r**anging. Radar is a device for locating objects in

space, direction finding and navigation by means of transmitted and reflected high-frequency radio waves.

radar imaging

The method of locating objects using radar. The direction of an object is ascertained by transmitting a beam of short-wavelength, short-pulse radio waves and picking up the reflected beam – its echo. Distance is determined by timing the journey of the radio waves (traveling at the speed of light) to the object and back again.

radiation

The emission of energy from a source in the form of electromagnetic waves or high-energy particles. The energy falls off as the inverse square of the distance from the source in the absence of absorption (*see* **inverse-square law**). The term is also applied to sound waves. The high-energy particles emitted in radioactivity are referred to as nuclear radiation. Radiation of heat is the transfer of heat by infrared rays. The heat can pass through a vacuum, travels at the same speed as light, can be reflected and refracted, and does not affect the medium through which it passes. For example, heat reaches the Earth from the Sun by radiation. Cosmic radiation consists of protons and other particles that reach the Earth from outer space. *See also* **ionizing radiation**.

CONNECTIONS

LIGHT AND THE SPECTRUM **106**

INVISIBLE RADIATIONS **118**

THE UNSTABLE ATOM **132**

---

radioactivity

The spontaneous emission of radiation from atomic nuclei. The process establishes an equilibrium in parts of the nuclei of unstable radioactive substances, allowing them to form a stable arrangement of nucleons (protons and neutrons) – that is, to form a non-radioactive (stable) element. This is most frequently accomplished by the emission of **alpha particles** (helium nuclei), **beta particles** (electrons and positrons), or **gamma radiation** (electromagnetic waves of very high frequency). The process takes place either directly, through a one-step decay, or indirectly, through a number of decays that transmute one element into another. This is called a decay series or chain and sometimes produces an element more radioactive than its predecessor.

In a radioactive atom, the instability of the particle arrangements in the nucleus (the ratio of neutrons to protons and/or the total number of both) determines the lengths of the half-lives of the isotopes of that atom, which can range from fractions of a second to billions of years. All **isotopes** of **relative atomic mass** 210 and greater are radioactive. Alpha, beta and gamma radiation all produce ionization.

radio telescope

An instrument for detecting radio waves from the Universe. Radio telescopes usually consist of a metal "dish" that collects and focuses radio waves the way a concave mirror collects and focuses light waves. Radio telescopes are much larger than optical telescopes, because the wavelengths they are detecting are much longer than the wavelength of light. Interferometry is a technique in which the output from two or more dishes is combined to give better resolution of detail than with a single dish.

radio waves

Electromagnetic radiation occurring in the frequency range from 10 kHz to more than 10 GHz and used to convey information. *See* **modulation**.

radius of curvature

The radius of a sphere of which a lens surface or a curved mirror forms a part. *See* **center of curvature**.

rainbow

An arc of spectral colors in the sky resulting from reflection and refraction of sunlight inside raindrops. *See* **spectrum**.

Raman effect

The scattering of white light into different wavelengths when it passes through a transparent medium, caused by interactions with the molecules of the medium.

Raoult's law

Dissolving a substance (solute) in a liquid (solvent) produces a solution whose vapor pressure is lowered below that of the pure solvent by an amount proportional to the mole fraction of the solute.

real image

See **image**.

rectifier

An electronic device used to obtain **direct current (DC)** from an alternating source of supply. The process is necessary because although amost all electrical power is generated, transmitted and supplied as **alternating current (AC)**, there are many devices, from television sets to amplifiers, that require direct current.

REFLECTION

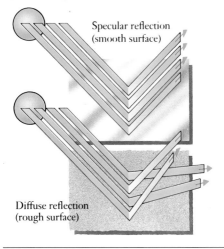

Specular reflection (smooth surface)

Diffuse reflection (rough surface)

reflection

The bouncing back or deflection of waves, such as light or sound waves, when they hit a surface. The laws of reflection are: **1** The angle of incidence (the angle between the incident ray and a perpendicular line drawn to the surface) is equal to the angle of reflection (the angle between the reflected ray and a line perpendicular to the surface). **2** The reflected ray is in the same plane as the incident ray and the normal to the reflecting surface at the point of incidence.

refraction

The bending of waves when they pass from one medium to another. It occurs because waves travel at different velocities in different media, which leads to a change in the direction of propagation in accordance with **Snell's law**. The laws of refraction are: **1** The incident ray, the refracted ray and the normal at the point of incidence are all in the same plane. **2** Snell's law.

CONNECTIONS

PRODUCING SOUND **98**

REFRACTION AND LENSES **112**

DISPERSION AND DIFFRACTION **114**

---

refractive index

A measure of the refraction of a ray of light as it passes between transparent media. If the angle of incidence is i and the angle of refraction is r, the refractive index is $n = \theta i / \theta r$. It is also equal to the speed of light in the first medium divided by the speed of light in the second and it varies with the wavelength of the light.

relative atomic mass

The mass of an atom relative to one-twelfth the mass of an atom of carbon-12. It depends on the number of **protons** and **neutrons** in the atom, the electrons having insignificant mass. If more than one isotope of the element is present, the relative atomic mass is calculated by taking an average that takes account of the relative proportions of each isotope, resulting in values that are not whole numbers. The term atomic weight, though commonly used, is not technically correct.

relative humidity

The pressure of water vapor in a gas (such as air) divided by the pressure the vapor would have if the gas were saturated with water, expressed as a percentage.

relative molecular mass

For a molecule of a substance, the sum of the relative atomic masses (atomic weights) of all the atoms in it. Also called molecular weight.

relativity

The theory of the relative rather than absolute character of motion and mass, and the interdependence of matter, time and space, as developed by the German physicist Albert Einstein in two phases. Special theory (1905) starts with the premises that **1** the laws of nature are the same for all observers in unaccelerated motion and **2** the speed of light is independent of the motion of its source. Einstein proposed that the time interval between two events was longer for an observer in whose frame of reference the events occur in different places than for the observer for whom they occur at the same place.

In the general theory of relativity (1915) the geometrical properties of space-time were conceived as modified locally by the presence of an object with mass. A planet's orbit around the Sun (as observed in three-dimensional space) arises from its natural trajectory in modified space-time; there is no need to invoke, as Isaac Newton did, a force of **gravity** coming from the Sun and acting on the planet. Einstein's theory predicted slight differences in the orbits of the planets from Newton's theory, which were observable in the case of Mercury. The new theory also said light rays should bend when they pass by a massive object, due to the object's effect on local space-time.

The predicted bending of starlight was observed during an eclipse of the Sun in 1919, when light from distant stars passing close to the Sun was not masked by sunlight. Einstein showed that for consistency with premises **1** and **2**, the principles of dynamics as established by Newton needed modification; the most celebrated new result was the equation $E = mc^2$, which expresses an equivalence between mass (m) and energy (E), c being the speed of light in a vacuum. Although it has since been modified in detail, the theory of general relativity remains central to modern astrophysics and cosmology; it predicts, for example, the possibility of black holes.

General relativity theory was inspired by the simple idea that it is impossible in a small region to distinguish between acceleration and gravitation effects (as in an elevator one feels heavier when the elevator accelerates upwards), but the mathematical development of the idea is highly sophisticated.

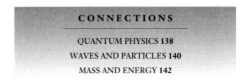

CONNECTIONS

QUANTUM PHYSICS 138

WAVES AND PARTICLES 140

MASS AND ENERGY 142

resistance

That property of a substance that restricts the flow of electricity through it, associated with the conversion of electrical energy to heat; also the magnitude of this property. Resistance depends on many factors, such as the nature of the material, its temperature, dimensions and thermal properties; degree of impurity; the nature and state of illumination of the surface; and the frequency and magnitude of the current. The SI unit of resistance is the ohm. *See* **Ohm's law**.

resonance

Phenomenon in which a system vibrates at its natural frequency by picking up vibrations (of the same frequency) from another source.

right-hand rule

If the directions of the thumb, first finger and second finger of the right hand are held at right angles, the thumb indicates the direction of movement, the first finger the direction of the magnetic field and the second finger the direction of induced current flow in electromagnetic induction.

safety valve

A valve, spring or dead-weight load fitted to a boiler or other pressure vessel, to allow steam or gas to escape to the atmosphere when the pressure exceeds the maximum safe value.

scanning

The systematic coverage of a defined area by a spotlight moving in series of progressive lines, such as a sweeping of the image area by an electron beam in a television picture tube, or by a directional radio antenna or a sonar beam.

Schrödinger equation

An equation that regards an electron in an atom as a three-dimensional stationary wave, whose solution gives the probability for the location of the electron. *See also* **uncertainty principle**.

secondary cell

A rechargeable electrical cell. The chemical reactions that occur in a secondary cell can be reversed by applying electricity and the cell can be restored to its original condition. Secondary cells are also called accumulators. *See* **battery** and **primary cell**.

secondary color

A color formed by mixing two **primary colors**: green, a secondary color, is a mixture of yellow and blue.

Seebeck effect

See **thermoelectricity**.

semiconductor

Any crystalline material with an electrical **conductivity** between that of metals (good) and insulators (poor). Semiconductors can be classified into two types, intrinsic and extrinsic. An intrinsic semiconductor has a high degree of chemical purity. Its conductivity is poor and depends largely on its temperature. Some common intrinsic semiconductors are single crystals of silicon, germanium and gallium arsenide.

Extrinsic semiconductors are intrinsic semiconductors in which conductivity has been improved by small additions of different substances (*see* **doping**). Silicon, for example, has poor conductivity at low temperatures, but this is greatly improved by doping with phosphorus. The process of doping gives such intrinsic semiconductors many more uses in high technology. An extrinsic semiconductor may be classified as *n*- or *p*-type depending on whether the impurity has an excess of negative charge (*n*-type) or a deficiency of negative charge (*p*-type). Semiconductor materials are widely used in transistors, rectifiers and integrated circuits (silicon chips).

CONNECTIONS

ELECTRONICS AND SEMICONDUCTORS 92

MINIATURE CIRCUITS 94

TELEVISION CAMERA 124

semipermeable membrane

A membrane that allows small molecules (such as those of water) to pass through it but does not allow the passage of large molecules (such as those of proteins).

SECONDARY CELL

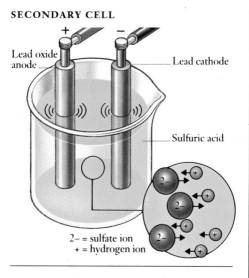

Lead oxide anode

Lead cathode

Sulfuric acid

2− = sulfate ion
+ = hydrogen ion

series circuit

An electrical circuit in which the components are connected end to end, so that the current flows through them sequentially. If the potential differences across two conductors of resistance R_1 and R_2, connected in series, are V_1 and V_2, respectively, then the ratio of those potential differences is given by the equation: $V_1/V_2 = R_1/R_2$. The total resistance R of those conductors is given by the formula $R = R_1 + R_2$. *See* **parallel circuit**.

silicon chip

An integrated circuit with microscopically small electrical components on a piece of silicon crystal only a few millimeters square. *See* **integrated circuit**.

simple harmonic motion

Oscillatory motion in which an object (or point) moves so that its acceleration toward a central point is proportional to its distance from it. A simple example is a pendulum, which also demonstrates another feature of simple harmonic motion: the maximum deflection is the same on each side of the central point. A graph of the varying distance over time is a sine curve, a characteristic of the oscillating current or voltage of an **alternating current (AC)**, which is another example of simple harmonic motion.

SI units

(In French, Système International d'Unités). The standard system of scientific units used by scientists worldwide. It is based on seven basic units: the meter (m) for length, kilogram (kg) for mass, second (s) for time, ampere (A) for electrical current, kelvin (K) for temperature, mole (mol) for amount of substance and candela (cd) for luminosity. There are numerous derived units (newton, joule, etc.) and scales of decimal multiples and submultiples, all with agreed symbols.

Snell's law

A wave refracted at a surface makes angles relative to the normal of the surface, written as $n_1\sin\theta_1 = n_2\sin\theta_2$, where n_1 and n_2 are the refractive indices on each side of the surface and θ_1 and θ_2 are the corresponding angles.

solenoid

A coil of wire, usually cylindrical, in which a magnetic field is created by passing an electric current through it (*see* **electromagnet**). This field can be used to move an iron rod placed on its axis. Mechanical valves attached to the rod can be operated by switching the current on or off, so converting electrical energy into mechanical energy.

solid

A state of matter that holds its own shape (as opposed to a liquid, which takes up the shape of its container, or a gas, which totally fills its container). According to **kinetic theory**, the atoms or molecules in a solid are not free to move but merely vibrate around fixed positions, such as those in crystal lattices.

solid-state physics

That branch of physics that deals with the study of the properties of solids, specifically the study of the electrical properties of semiconductors and their electronic structure. Solid-state devices are electronic components which consist entirely of solids, such as semiconductors, transistors, etc., without heating elements, as in thermionic tubes (valves). Solid-state physics also covers photoconductivity and superconductivity. The term condensed-matter physics has recently been introduced to cover the study of crystalline solids, amorphous solids and liquids.

sonar

An acronym for **so**und **na**vigation and **r**anging. Sonar is a method of locating underwater objects by the reflection of ultrasonic waves (echoes). The time taken for an acoustic beam to travel to the object and back to the source enables the distance to be found because the velocity of sound in water is known. The process is similar in principle to that used in **radar**.

sound

A physiological sensation received by the ear, originating in a vibration (pressure variation in the air) that communicates itself to the air and travels in every direction, spreading out as an expanding sphere. All sound waves in air travel with a speed dependent on the temperature; under ordinary conditions, this is about 330 m/sec (1070 ft/sec). The loudness of a sound is dependent primarily on the amplitude of the vibration of the air. The lowest note audible to an adult human has a frequency of about 20 Hz and the highest one of about 15,000 Hz. *See* **pitch**.

CONNECTIONS

SOUND ENERGY 96

PRODUCING SOUND 98

SPEED OF SOUND 100

ULTRASOUND 102

SOUND RECORDING 104

---

sound barrier

A hypothetical speed limit to flight through the atmosphere at around 1170 km/h (727 mph), because an incorrectly designed aircraft suffers severe buffeting at near-sonic speed owing to the formation of shock waves, causing the wings to fail in the early aircraft that attempted it. The sound barrier was first broken in 1947.

spark plug

A device that produces an electric spark in the cylinder of a gasoline engine to ignite the fuel mixture. It consists of two electrodes insulated from one another. High-voltage electricity is fed to a central electrode via the distributor. At the base of the electrode, inside the cylinder, the electricity jumps to another electrode grounded to the engine body, creating a spark.

specific gravity

An alternative term for relative **density**.

specific heat capacity

The quantity of heat required to raise unit mass (1 kg) of a substance by 1 K. The unit of specific heat capacity in the SI system is the joule per kilogram per kelvin ($J\ kg^{-1}\ K^{-1}$).

spectrometer

An instrument used for studying the composition of light. The range, or spectrum, of wavelengths emitted by a source depends upon its constituent elements and may be used to determine its chemical composition. The simpler forms of spectrometer analyze only visible light. A collimator receives the incoming rays and produces a parallel beam, which is then split into a spectrum by a diffraction grating or prism mounted on a turntable. As the turntable is rotated each of the constituent colors of the beam may be seen through a telescope, and the angle at which each has been deviated may be measured on a circular scale. From this information the wavelengths of the colors of light can be calculated.

spectrum

An arrangement of frequencies or wavelengths that occurs when electromagnetic radiations are separated into their constituent parts. Visible light is part of the electromagnetic spectrum and most sources emit waves over a range of wavelengths that can be broken up or "dispersed"; white light can be separated through a prism into red, orange, yellow, green, blue, indigo and violet. *See* **color**.

speed

The rate at which an object moves. Speed in kilometers per hour is calculated by dividing the distance traveled in kilometers by the time taken in hours. Speed is a scalar quantity, because the direction of motion is not specified. This makes it different from velocity (speed in a particular direction), which is a vector quantity.

speed of light

The speed at which light and other electromagnetic waves travel through empty space. Its value is 299,792,458 m/sec (186,281 miles/sec). According to Einstein's theory of relativity, the speed of light is the highest speed possible, and it is independent of the motion of its source and of the observer. It is

SIMPLE HARMONIC MOTION

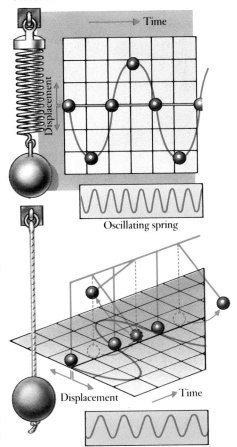

Oscillating spring

Swinging pendulum

impossible to accelerate any material object to this speed because it would require an infinite amount of energy.

speed of sound
In dry air at 0°C at sea level, a speed of 331.7 meters per second. It increases with temperature and with the density of the medium through which it travels.

spherical aberration
A defect of an optical instrument in which the image is blurred because different parts of a spherical lens or mirror have slightly different focal lengths.

standard temperature and pressure
(abbreviated as STP) A temperature of 273.15 K (0°C) and a pressure of 101,325 pascals (760 millimeters of mercury).

static electricity
The charge acquired by an object by means of electrostatic induction or friction. Rubbing various materials can produce static electricity, seen in the sparks produced on combing one's hair or removing a nylon shirt. In some processes static electricity is useful, as in paint spraying where the parts to be sprayed are charged with electricity of opposite polarity to that on the paint droplets.

CONNECTIONS

ELECTRICITY AND MAGNETISM **74**
STATIC ELECTRICITY **78**
ELECTRICITY AND OTHER ENERGY **90**

---

statics
A branch of mechanics that is concerned with the behavior of objects at rest and forces in equilibrium. Moving objects are studied in **dynamics.**

STRESS

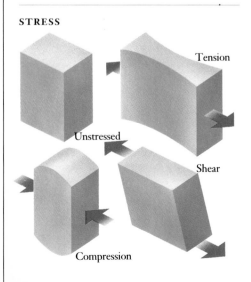

Tension

Unstressed

Shear

Compression

steam turbine
An engine in which the pressure of steam is made to spin a rotating shaft (rotor) inside a fixed casing (stator) by pushing on sets of angled vanes attached to the shaft. This type of steam turbine is known as the reaction turbine. Less commonly used is the impulse turbine which works by directing a jet of steam at the vanes. Steam turbines are used to drive large electricity generators in power stations. *See also* **gas turbine** and **internal-combustion engine**.

stereophonic sound
A system of sound reproduction that uses two complementary channels leading to two loudspeakers, which gives a more natural depth to the sound than with a single channel. Stereo recording began with the introduction of two-track magnetic tape in the 1950s.

strain
A measure of the distortion of an object when a deforming force (stress) is applied to it. Strain is a ratio of the extension or compression of the object (its length, area, or volume) to its original dimensions. For example, linear strain is the ratio of the change in length of an object to its original length. *See* **Hooke's law**.

streamline flow
A type of fluid (gas or liquid) flow in which there is no turbulence and the particles of the fluid follow continuous paths, either at constant velocity or at a velocity that alters in a predictable way.

stimulated emission
The principle of the **laser**, in which a photon (quantum of light) causes an electron in an atom to occupy a lower energy level and emit another photon.

stress
The force per unit area on an object which causes it to deform. It is a measure of the internal forces between the particles of the material as they resist separation, compression or sliding in response to externally applied forces. *See* **Hooke's law**.

subatomic particle
Any particle that is smaller than an atom. Such particles may be indivisible **elementary particles**, such as the electron and quark, or they may be composites, such as the proton, neutron and alpha particle.

sublimation
A process in which a heated solid changes directly into a vapor without first melting.

subsonic speed
A speed that is less than Mach 1 – the speed at which sound travels. *See* **Mach number**, **sound barrier** and **supersonic speed**.

superconductivity
The ability of certain materials to conduct electricity without resistance at very low temperatures. The resistance of some metals and metallic compounds decreases uniformly with decreasing temperature until at the superconducting point, within a few degrees of absolute zero, the resistance suddenly falls to zero. *See also* **semiconductor** and **superfluidity**.

superfluidity
The property of liquid helium that enables it to flow without friction at very low temperatures. There is a basic connection between superconductivity and superfluidity, so that sometimes a superconductor is called a charged superfluid.

supersonic speed
A speed greater than that at which sound travels, measured in Mach numbers. In dry air at 0°C (32°F), sound travels at about 1170 km/h (727 mph) at sea level, but its speed decreases with altitude until, at 12,000 m (39,000 ft), it is only 1060 km/h (658 mph). *See* **sound barrier**.

surface tension
The property of a liquid that makes it behave as though its surface is enclosed in a weak elastic skin. Surface tension results from the fact that the exposed surface of a liquid has a tendency to contract to the smallest possible area. This is due to unequal cohesive forces between molecules at the surface as compared with molecules in the interior of the liquid.

susceptibility
A measure of the extent to which a material is magnetized by an applied magnetic field. Expressed mathematically, it is the ratio of the intensity of magnetization produced in a material to the intensity of the magnetic field to which the material is exposed. Diamagnetic materials have small negative susceptibilities; paramagnetic materials have small positive susceptibilities; ferromagnetic materials have large positive susceptibilities. *See* **magnetism**.

synchrotron
See **accelerator**.

telescope
See **radio telescope** and **astronomical telescope**.

television

The transmission, reception, and reproduction of visual images by radio waves. For transmission, a television camera converts the pattern of light it takes in into a pattern of electrical charges. This is scanned line by line by a beam of electrons from an electron gun, resulting in variable electrical signals that represent the visual picture. These vision signals are combined with a radio carrier wave and broadcast as electromagnetic waves. The TV aerial picks up the wave and feeds it to the receiver (TV set). This separates out the vision signals, which pass to a cathode-ray tube. The vision signals control the strength of a beam of electrons from an electron gun, aimed at the screen and making it glow more or less brightly. At the same time the beam is made to scan across the screen line by line, mirroring the action of the gun in the TV camera. The result is a recreation of the pattern of light that entered the camera. Thirty pictures are built up each second with interlaced scanning in North America (25 in Europe), with a total of 525 lines in North America and Japan (625 lines in Europe), although current developments in high-definition TV (HDTV) are leading to the introduction of over 1200 lines.

CONNECTIONS

ELECTRONICS AND SEMICONDUCTORS 92
TELEVISION CAMERA 124
TELEVISION RECEIVER 126

temperature

The state of hotness or coldness of an object and the condition that determines whether or not it will transfer heat to, or receive heat from, another object according to the laws of thermodynamics: heat flows from a higher temperature to a lower. It is measured using the **Celsius, Kelvin** or **Fahrenheit** scales. To convert degrees Celsius to degrees Fahrenheit, multiply by $\frac{9}{5}$ and add 32; Fahrenheit to Celsius, subtract 32, then multiply by $\frac{5}{9}$. A useful quick approximation for converting Celsius to Fahrenheit is to double the Celsius and add 30; for example, 12°C = 24 + 30 = 54°F (the precise value is 53.6°F). *See also* **absolute temperature, boiling point, melting point** and **triple point.**

CONNECTIONS

HEAT ENERGY 68
HEATING AND COOLING 70
MEASURING AND USING HEAT 72
NUCLEAR FUSION 136

terminal velocity

For an object falling through a gas or liquid because of gravity, the constant velocity (speed in a particular direction) it reaches when there is no resultant force acting on it.

thermal capacity

The amount of heat required to raise the temperature of a system by 1 K. The SI unit is JK^{-1}. *See* **heat capacity**.

thermal neutron

A slow-moving neutron that can be captured by an atomic nucleus (and possibly initiate **nuclear fission**).

thermionic emission

The emission of electrons, usually into a vacuum, from a heated conductor. It is the basis of thermionic tubes (valves) and the **electron gun** in cathode-ray tubes.

thermocouple

An electric temperature-measuring device consisting of a circuit with two wires made of different metals welded together at their ends. A current flows in the circuit when the two junctions are maintained at different temperatures (Seebeck effect). The electromotive force generated (measured by a millivoltmeter) is proportional to the temperature difference.

thermodynamics

The branch of physics that deals with the transformation of heat into and from other forms of energy. It is the basis of the study of the efficient working of engines, such as the steam and internal-combustion engines. The three laws of thermodynamics may be stated as follows: **1** energy can be neither created nor destroyed, heat and mechanical work being mutually convertible; **2** it is impossible to convey heat from one body to another at a higher temperature without work being done; **3** it is impossible by any procedure, no matter how idealized, to reduce any system to the absolute zero of temperature in a finite number of operations. There is also a "zeroth law" which states that two systems each in equilibrium with a third system are also in equilibrium with each other. Put into mathematical form, these laws have widespread applications in physics.

thermoelectricity

An electric current generated by a temperature difference. An electromotive force is generated in a circuit containing two different metals or semiconductors when their junctions are maintained at different temperatures. Known as the Seebeck effect, this is the basis of the thermocouple.

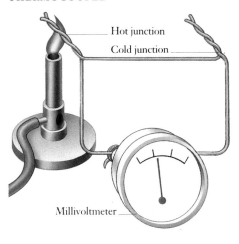

Hot junction
Cold junction
Millivoltmeter

thermometer

An instrument for measuring temperature. There are many types, designed to measure different temperature ranges to varying degrees of accuracy. Each makes use of a different physical effect of temperature. Expansion of a liquid is employed in common liquid-in-glass thermometers, such as those containing mercury or dyed alcohol. The more accurate gas thermometer uses the effect of temperature on the pressure of a gas held at constant volume. A resistance thermometer takes advantage of the change in the electrical resistance of a conductor (such as a platinum wire) as its temperature changes. Another electrical thermometer is the thermocouple. Mechanically, changes in temperature can be indicated by the change in curvature of a bimetallic strip (as commonly used in a thermostat).

thermonuclear reaction
See **nuclear fusion**.

thermostat

A device that automatically controls temperature, such as that of an electric heater (which the thermostat switches on and off to maintain a constant temperature).

time dilation

The term given to the phenomenon first proposed by Albert Einstein, in which the time interval between two events appears to be longer when they occur in a frame of reference that is in motion relative to the observer's frame of reference, than when they occur at rest in relation to the observer. Time dilation is a consequence of the special theory of **relativity**.

torque

A turning moment about an axis produced by a force acting at right angles to a radius.

total internal reflection

The reflection of a light ray as it reaches the edge of a transparent medium in which it is traveling. *See* **critical angle**.

transceiver

A piece of telecommunication equipment in which the electrical circuitry is used for both transmission and reception.

transducer

A power-transforming device that enables **energy** in any form (electrical, acoustical or mechanical) to flow from one system to another. The energy flowing to and from a transducer may be of the same or of different forms. For example, an electric motor receives electrical energy and delivers a mechanical output; a pick-up (needle) crystal receives mechanical energy from the stylus of a record player and delivers an electrical output; and a loudspeaker receives an electrical input and delivers an acoustical output.

transformer

A device in which an alternating current of one voltage is transformed by electromagnetic induction to another voltage without a change of frequency. A transformer has two coils, a primary for the input and a secondary for the output, wound on a common iron core. The ratio of the primary to the secondary voltages is directly proportional to the number of turns in the primary and secondary coils; the associated currents are inversely proportional to the numbers of turns.

transistor

A solid-state electronic component made of semiconductor material, with three or more electrodes, that can regulate a current passing through it. A transistor can act as an amplifier, oscillator, photocell or switch, and operates on a very small amount of power. Transistors commonly consist of a tiny sandwich of germanium or silicon, so that its alternate layers have different electrical properties.

triode

A three-electrode thermionic tube (valve) containing an anode (plate) and a cathode (as does a diode) with an additional electrode called a grid. Small variations in voltage on the grid result in large variations in the current. The triode was commonly used in amplifiers but has now been almost entirely superseded by the **transistor**.

triple point

The temperature and pressure at which the vapor, liquid and solid phases of a substance are in equilibrium. For water the triple point is 273.16 K and 610 N m^{-2}. This point forms the basis of the definition of the **Kelvin** temperature scale.

tritium

A rare isotope of hydrogen in which the nucleus contains two neutrons and one proton – that is, it has a mass number of 3. It has a half-life of about 12.5 years. Its abundance in natural hydrogen is one atom in 107.

turbulence

A form of fluid flow in which the particles of the fluid move in an irregular eddying motion, resulting in an exchange of momentum from one part of the fluid to another. *See also* **airfoil** and **streamline flow**.

ultrasound

Pressure waves similar to sound waves but occurring at frequencies above 20,000 Hz.

ultrasonic scanning

The use of ultrasound to produce echoes as a means of measuring the depth of the sea in a similar way to sonar or to detect flaws in metal. In medicine, ultrasound is used to investigate various body organs.

ultraviolet radiation

Electromagnetic radiation invisible to the human eye, of wavelengths from about 400 to 4 nm (where the X-ray range begins). The radiation may be detected with ordinary photographic plates or films. It can also be studied by its fluorescent effect on certain materials.

uncertainty principle

The principle in quantum mechanics that it is meaningless to speak of a particle's position, momentum or other parameters except as results of measurements. Measuring, however, involves an interaction (such as a

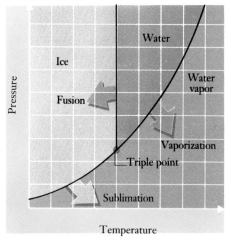

TRIPLE POINT

photon of light bouncing off the particle under scrutiny), which must disturb the particle, though the disturbance is noticeable only at an atomic scale. The principle implies that one cannot, even in theory, predict the moment-to-moment behavior of such a system and that it is impossible to determine the position and momentum of a particle simultaneously. The principle was first stated in 1927 by the German physicist Werner Heisenberg.

vacuum

In general, a region completely empty of matter; in physics, any enclosure in which the gas pressure is considerably less than atmospheric pressure (101,325 pascals).

vacuum tube

An electronic device, also called an electron tube or valve, that consists of a glass tube containing gas at low pressure, which is used to control the flow of electricity in a circuit. Three or more metal electrodes are inset into the tube. By varying the voltage on one of them, called the grid, the current through the valve can be controlled and the valve can act as an amplifier. Vacuum tubes have been replaced by transistors for most applications.

vapor

A state of matter similar to gas. The molecules in a vapor move randomly and are far apart, the distance between them – and therefore the volume of the vapor – being limited only by the walls of any vessel in which they might be contained. A vapor differs from a gas only in that a vapor can be liquefied by increased pressure, whereas a gas cannot unless its temperature is lowered below its critical temperature; it then becomes a vapor and may be liquefied.

velocity

The speed of an object in a given direction. Velocity is a vector quantity, because it is defined by its direction as well as its magnitude (or speed, which is a scalar quantity). If the direction of motion of an object changes, even if it is traveling at constant speed, then its velocity also changes and it therefore accelerates. The velocity at any instant of a particle traveling in a curved path is in the direction of the tangent to the path at the instant considered.

velocity ratio

See **distance ratio**.

video

The general term used to describe the electronic handling of visual images, usually accompanied by sound. A video camera (camcorder) is a portable television camera that takes moving pictures electronically on magnetic tape. The output is recorded on video cassette and is played back on a television screen via a video cassette recorder (VCR). The picture information is stored as a line of varying magnetism, or track, on a plastic tape covered with magnetic material. The main difficulty – the huge amount of information needed to reproduce a picture – is overcome by arranging the video track diagonally across the tape. During recording, the tape is wrapped around a drum in a spiral. The recording head rotates inside the drum. The combination of the forward motion of the tape and the rotation of the head produces a diagonal track. The audio signal accompanying the video signal is recorded as a separate track along the edge of the tape.

CONNECTIONS

TELEVISION CAMERA **124**

TELEVISION RECEIVER **126**

---

virtual image
See **image**.

viscosity
The resistance of a fluid to flow, caused by internal friction, which makes it resist flowing past a solid surface or other layers of the fluid. It applies to the motion of an object through a fluid and to the motion of a fluid passing by an object or through a pipe.

visible light
Electromagnetic radiation to which the human eye is sensitive. Visible light is in the wavelength range of 700 to 400 nm.

voltage
The **potential difference** between two points in an **electrical circuit** when current flows through a **resistance**. The SI unit is the volt. *See* **Ohm's law**.

voltmeter
An instrument for measuring potential difference (voltage), which is connected in parallel with the component across which the measurement is to be made. A common type is based on a sensitive current-detecting moving-coil galvanometer in series with a high-value resistor (multiplier). To measure an alternating current (AC) voltage, the circuit must usually include a rectifier.

water turbine
A turbine that works on the same principle as a **steam turbine**, but uses water instead of steam to drive the rotating shaft.

WAVE

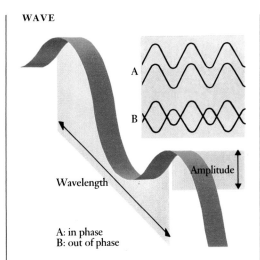

A: in phase
B: out of phase

Wavelength
Amplitude

wave
A disturbance traveling through a medium (or space). In a longitudinal wave (such as a sound wave), the disturbance is parallel to the wave's direction of travel; in a transverse wave (such as an electromagnetic wave), it is perpendicular. The chief characteristics of a wave are its speed, wavelength, frequency and amplitude. Waves transport energy, and show the properties of reflection, refraction, diffraction and interference. The speed c, frequency f and wavelength λ of a wave are related by the equation $c = f\lambda$. *See also* **wave mechanics** and **wave number**.

CONNECTIONS

SOUND ENERGY **96**

ULTRASOUND **102**

LIGHT AND THE SPECTRUM **106**

INVISIBLE RADIATIONS **118**

RADIO WAVES **120**

WAVES AND PARTICLES **140**

---

wavelength
The distance between successive points of equal phase of a wave. The wavelength of a **light** wave determines its **color**; for instance, red light has a wavelength of about 700 nm. The complete range of wavelengths of **electromagnetic waves** is called the **electromagnetic spectrum**.

wave mechanics
The modern form of **quantum theory** in which events on an atomic or subatomic scale are considered in terms of the interaction between wave systems, as expressed by the **Schrödinger equation**. This equation describes the wavelike properties of a particle and associates a wave function with the particle's probable location. *See also* **orbital** and **uncertainty principle**.

wave number
The reciprocal of the wavelength of electromagnetic radiation – that is, the number of waves in a unit distance.

weight
The force exerted on an object by gravity. It depends on its mass (the amount of material in it) and the strength of the Earth's gravitational pull, which decreases with height. Consequently, an object weighs slightly less at the top of a mountain than at sea level. On the Moon, an object has only one-sixth of its weight on Earth, because the pull of the Moon's gravity is one-sixth that of the Earth. If the mass of an object is m kilograms and the gravitational field strength is g newtons per kilogram, its weight W in newtons is given by: $W = mg$.

weightlessness
Property of an object that is in free fall or not in any gravitational field.

work
A measure of energy transferred from one system to another to cause an object to move. Work is not the same as energy (the capacity to do work, which is also measured in joules) or power (the rate of doing work, measured in watts – joules per second). Work W is equal to the product of the force F used and the distance d moved by the object in the direction of that force: $W = Fd$.

X rays
A form of short-wavelength electromagnetic radiation (10^{-3} to 10 nm) produced when high-energy electrons from a heated filament cathode strike the surface of a target (usually made of tungsten) on the face of a massive heat-conducting anode, between which a high alternating voltage (about 100 kV) is applied. Electrons passing near a nucleus in the target are accelerated and so emit a continuous form of radiation ranging up from a minimum wavelength. In addition, the electrons may eject an electron from the inner shell of one of the target atoms and the subsequent transition of an electron of higher energy to this level produces radiation of specific wavelength. This characteristic X-ray spectrum is specific for the target material. X rays may be detected by photographic plates or by a counting instrument. *See also* **gamma rays**.

yield point
The stress beyond which a material deforms by a relatively large amount for a small increase in stretching force. Beyond this stress, the material no longer obeys **Hooke's law**. *See* **elastic limit**.

1

PROPERTIES
of Matter

EVERYTHING IN THE WORLD is made up of matter. All forms of matter are, in turn, made up of tiny particles called atoms, or of combinations of atoms called molecules. Much too small to be seen even with the most powerful optical microscopes, atoms and molecules account for the physical forms that matter may take, and for the way matter behaves in each of its various forms.

All matter exists in one of three states: solid, liquid or gas. Rocks are solids: they have a definite shape and obvious mass. The oceans are mainly a liquid – water. Liquids also have mass: a full bottle of water is heavier than an empty one. Like other liquids, water has no shape of its own and takes on the shape of its container. Air is a mixture of different gases, which make up a third kind of matter. Insubstantial and lightweight, a gas must be kept in a closed container or it escapes, but it still has mass. Many substances can exist in more than one state. For instance, water is normally a liquid, but when it freezes it becomes solid ice.

An unusual fourth kind of matter occurs in the Sun and other stars. Known as a plasma, it consists of fragments of atoms, and can be produced on Earth only in experimental nuclear fusion reactors or by a hydrogen bomb explosion.

Snow monkeys bask in the warm waters of a lake created by hot springs in the chilly mountains of northern Japan. Surrounded by clouds of steam (gaseous water vapor), they are also in contact with two other states of matter: liquid water and snow, which is water in its solid state. Like many other substances, water can coexist in all three phases of matter simultaneously.

GASES AND VAPORS

Gas, or vapor, is the simplest form of matter. It has no structure, consisting of large numbers of independent particles (atoms or molecules). The individual particles have mass, which means that some gases are heavier than others. Because they are continually moving, the atoms or molecules possess kinetic energy – the energy of movement. The amount of kinetic energy depends mainly on how fast they are moving, which in turn depends on the temperature of the gas: the hotter it is, the faster the molecules move around. Kinetic theory explains many phenomena connected with gases.

Air, the gas with which we are most familiar, has no container to hold it, and the force of gravity keeps it near the surface of the Earth. Gas pressure is a powerful natural force. The force or pressure of the gases in the atmosphere derives from the weight of all the molecules above, which press on every square centimeter of every object in it. Atmospheric pressure is not constant: it falls with altitude. Weather results from occasional local variations in pressure and temperature. If the air molecules are heated the pressure falls and the hot air rises. Surrounding air masses rush in to replace the warm air, creating wind.

Most gases are studied in a container. The moving particles of a gas collide with one another and with the walls of the container. Particles striking the walls exert a force, and the effect of all these collisions accounts for the pressure that a gas exerts on its container.

If extra pressure is applied to a gas its volume decreases, and the new volume is inversely proportional to the pressure (as long as the temperature does not change). This is known as Boyle's law (1662). But if the pressure of the gas is kept constant and it is heated, its volume increases, a relationship named Charles' law (1787). These laws can be stated mathematically. If the volume of the gas is

▶ Hot air balloons put the physical laws relating to gases to good use. As the air in the balloon is heated by a burner underneath, its volume increases and becomes less dense than the colder air surrounding the balloon. As a result, the balloon rises. Once aloft, it is blown around on the air currents caused by local weather conditions. A study of the physics of gases explains this phenomenon precisely, and allows the balloon to be designed and flown safely.

▶ As a maple seed falls through the air, pressure differences on the upper and lower surfaces of the "wings" of the seed cause it to describe a spiral motion through the air. A similar principle enables a helicopter to fly.

Envelope

Gas burner

Control valve

Bottles of fuel gas

Pointer

Hairspring

Evacuated chamber

Spring

called V, its pressure p and its absolute temperature (temperature above absolute zero) T, Boyle's law can be stated "pV = constant", and Charles', "V/T = constant". From these, a third gas law can be deduced: "p/T = constant" (at constant volume, pressure is proportional to temperature). These laws explain why the balloon rises when the air inside is heated.

Kinetic theory explains other gas phenomena. Gas in a porous container loses particles gradually through the tiny holes in the material. Lighter gases have atoms or molecules with higher kinetic energy – they move faster and diffuse through the walls more quickly. The mid-19th century Scottish physicist Thomas Graham discovered that the rate at which a gas diffuses in this way is inversely proportional to the square root of its density. This principle has important applications. For example, in the nuclear industry, the two chief isotopes (molecular variants) of the element uranium must be separated to "enrich" the material sufficiently for use as a nuclear fuel. This can be done by making them into a gaseous compound and letting it diffuse through a number of filters. The lighter isotope is less dense, so it diffuses faster.

The English physicist John Dalton made another important discovery in 1801. If two or more gases are mixed, each exerts a pressure equal to the pressure it exerts on its own in the container. The particles of the individual gases collide but do not interfere with each other, and keep striking the walls of the container to exert pressure.

In 1811 the Italian physicist Amadeo Avogadro put forward the theory that, at the same temperature and pressure, equal volumes of all gases contain the same number of particles. The number of particles in a gram molecular weight of a light element (2 grams in the case of hydrogen) or of a heavier molecule such as carbon dioxide (44 grams) is identical. The number – just over 600,000 billion billion – is called Avogadro's constant. It is used to calculate the quantities of raw materials for chemical reactions.

◄ **A barometer is used to measure atmospheric pressure. An aneroid type is based on a metal chamber containing a vacuum, which changes shape with changes in air pressure. Movements of the chamber cause the pointer to move on the dial.**

USING GAS PRESSURE

THE sails of ships and windmills were among the first human inventions to harness the power of gases under pressure – in this case, the pressure of gases in the atmosphere. Atmospheric pressure is also used in a simple lift pump for raising water from a well, acting on the surface of the water to lift it up. Other compressed gases, stored in metal cylinders or pumped from a container, can power equipment from pneumatic drills and jackhammers to aerosols such as perfume sprays.

Another use of gas under pressure sparked off the Industrial Revolution and was the key invention in the beginnings of modern technology. This was the steam engine. It was invented in the early 1700s and served as the major power source for industry and transport for more than 100 years, until it was replaced by electric motors and internal combustion engines. These newer sources of power themselves make use of gas pressure: turbines are used to drive the generators in power stations, while the pistons of an internal-combustion engine are powered by the expansion of the fuel gases after ignition.

The story of the steam engine began with Thomas Savery, an English engineer who in 1696 made a pump that employed a combination of steam and atmospheric pressure to pump water from mines. It relied on the fact that steam exerts a pressure but leaves a vacuum when condensed to form water. Sixteen years later Thomas Newcomen, from Cornwall in southwest Britain, built an engine that employed a cylinder and piston arrangement to transfer power to ordinary pumps. Then, beginning in 1769, James Watt designed steam engines that could drive industrial machinery more strongly and reliably than earlier water- or windmills. The new engines were also made into rail locomotives and traction engines for agriculture. These used the pressure of steam to drive a piston in a cylinder back and forth; the piston's motion was then transferred by linking rods to wheels.

The application of steam power to ocean-going ships brought an additional transportation revolution, freeing sailors from the uncertainties of the winds and ocean currents. Many marine engines took maximum advantage of the steam by using it at high-pressure in a small cylinder, then reusing the exhaust gases in a medium-pressure cylinder, and a third time at a lower pressure still. Steam power is also the principal driving force for modern electric power stations. Steam is raised in a boiler, which heats water by burning coal, oil or natural gas, or using heat from a nuclear reactor. Steam pressure spins the blades of turbines, and again three or more stages of gradually reduced pressure are employed. Regardless of what is used – fossil fuels or nuclear energy – large-scale electricity generation in most parts of the world is likely to continue to rely on steam-driven turbines.

Gas turbines can be used to power small-scale alternators. These work on the same principle as a steam turbine, but the energy comes from the gases that result from burning a fuel such as kerosene. Gas turbines find their main application, however, as the jet engines used to power aircraft. Air is compressed by the fans at the front of the engine and forced into the combustion chamber where ignition takes place. The exhaust gases are forced out of the rear in a powerful jet, creating a forward thrust. One advantage of the jet engine is that its performance improves with speed.

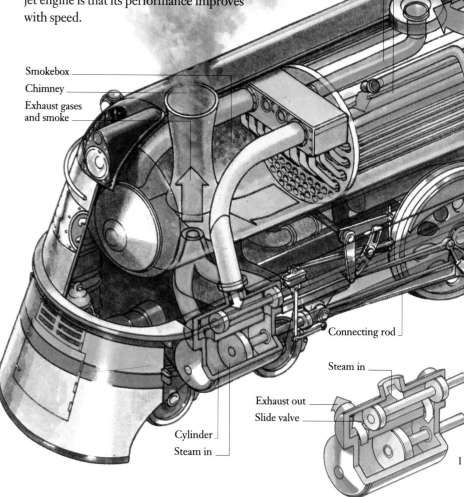

▼ A steam locomotive burns fuel such as coal, oil or wood to boil water to make the steam that drives its engine. The fuel is burned in the firebox, and the hot gases pass along firetubes that run the length of the boiler, surrounded by water. Steam collects at the top of the boiler and passes to the cylinders. The pressure of the steam moves pistons back and forth. Exhaust steam forces smoke out of the chimney and "sucks" hot gases through the firetubes.

Boiler
Steam
Smokebox
Chimney
Exhaust gases and smoke
Connecting rod
Steam in
Exhaust out
Slide valve
Cylinder
Steam in

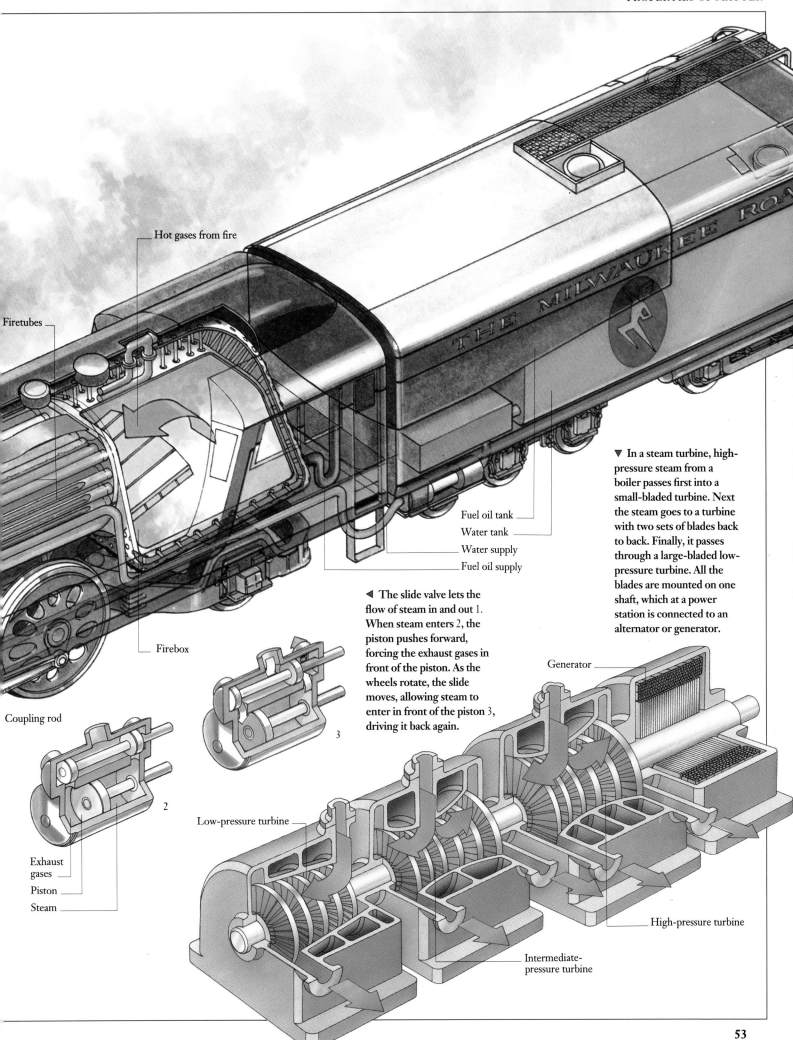

Hot gases from fire

Firetubes

Firebox

Coupling rod

Exhaust gases

Piston

Steam

2

3

Fuel oil tank

Water tank

Water supply

Fuel oil supply

◀ The slide valve lets the flow of steam in and out 1. When steam enters 2, the piston pushes forward, forcing the exhaust gases in front of the piston. As the wheels rotate, the slide moves, allowing steam to enter in front of the piston 3, driving it back again.

▼ In a steam turbine, high-pressure steam from a boiler passes first into a small-bladed turbine. Next the steam goes to a turbine with two sets of blades back to back. Finally, it passes through a large-bladed low-pressure turbine. All the blades are mounted on one shaft, which at a power station is connected to an alternator or generator.

Generator

Low-pressure turbine

Intermediate-pressure turbine

High-pressure turbine

THE LIQUID STATE

OIL and water are both liquids, but their physical properties are different. They can flow, yet their molecules are held together strongly enough by cohesive forces to take up a definite amount of space, though they have no shape. Both exert pressure, which for a liquid depends on its density and its depth. The liquid pressure occurring at the bottom of the sea is hundreds of times greater than at the surface.

At a particular point in a container of liquid, the pressure is the same in all directions. One special property of liquid is that, unlike a gas, it cannot be compressed. When pressure is applied at one point, it is immediately applied throughout the liquid: toothpaste, a thick liquid, squirts from the tube whether it is squeezed in the middle or at the end. This is an illustration of the principle of hydraulic mechanisms, which use the pressure caused by a relatively small force on a small piston to produce a very large force on a larger piston. Using a hydraulic jack, a person can single-handedly pump up enough pressure to lift a heavy truck.

Oil floats on water because oil is less dense. However, oil is thicker, or more viscous, than water, because of the greater attraction between its molecules. In a thick liquid such as oil or molasses, the molecules are strongly attracted to each other. They do not slide past each other easily, and the liquid is relatively difficult to pour.

Heating a viscous liquid decreases the molecular attraction; the liquid thins and becomes easier to pour. Increasing the pressure on a liquid forces the molecules closer together and increases viscosity. This is an important property of lubricating oils, which become more viscous with the high pressures between sliding parts and moving gears. If this were not so, the oil would be squeezed out and lubrication would fail, causing the parts to grind together.

Other physical properties of liquids may be explained by the cohesive force between molecules. The force between molecules at the surface, for example, creates the effect of a "skin" on the liquid. This is surface tension, which makes raindrops form spheres, holds soap bubbles together, and allows small insects to walk on water. Surface tension also makes it possible to float a small needle on water. Adding detergent to water with a floating needle causes the needle to sink, because detergent lowers surface

tension. A liquid with low surface tension rises up a narrow capillary tube, which is why a porous material such as a sponge or tissue paper soaks up water. If a capillary tube is placed in a liquid that has a high surface tension, such as mercury, the level of the liquid in the tube falls.

At the surface of a liquid, some of the vibrating molecules escape from the surface. This is the process of evaporation, and it can be increased by raising the temperature of the liquid. At a sufficiently high temperature the liquid boils, and molecules rapidly leave the surface to form a gas or vapor.

Decreasing the pressure on a liquid makes this process easier, reducing the boiling point. This is why water boils at a lower temperature at the top of a mountain (where atmospheric pressure is lower) than at sea level. Increasing the pressure on a liquid raises its boiling point, which is the principle of the pressure cooker, a widely used kitchen appliance.

▲ In drilling an oil well, liquid mud lubricates the drill as it bores through rock. When the drill is removed, pressure forces the mud upward. Workers then cap off the bore.

◄ A bug walks on water, its weight supported by the "skin" created by surface tension. This is possible because the water molecules at the surface are attracted to each other by a force of cohesion stronger than the force of attraction between surface molecules and the ones below them in the bulk of the liquid. Cohesion also makes the surface of the water turn up to form a meniscus at the edge of its container. These two effects work together to make water and similar liquids rise up a narrow capillary tube.

▼ Hydraulic machinery makes use of the fact that a liquid cannot be compressed. As a result, pressure applied at any point in a liquid is transferred equally throughout it. The hydraulics in the digger arm BELOW are filled with a type of oil, pressurized by a rotary pump. Valves operated by a cable controlled by the lever direct the high-pressure oil to one or the other side of a piston inside the cylinder, producing movement of the linkage – and hence the arm – in the direction required. Three such hydraulic cylinders are needed to produce the arm's full range of movements and to manipulate the scoop.

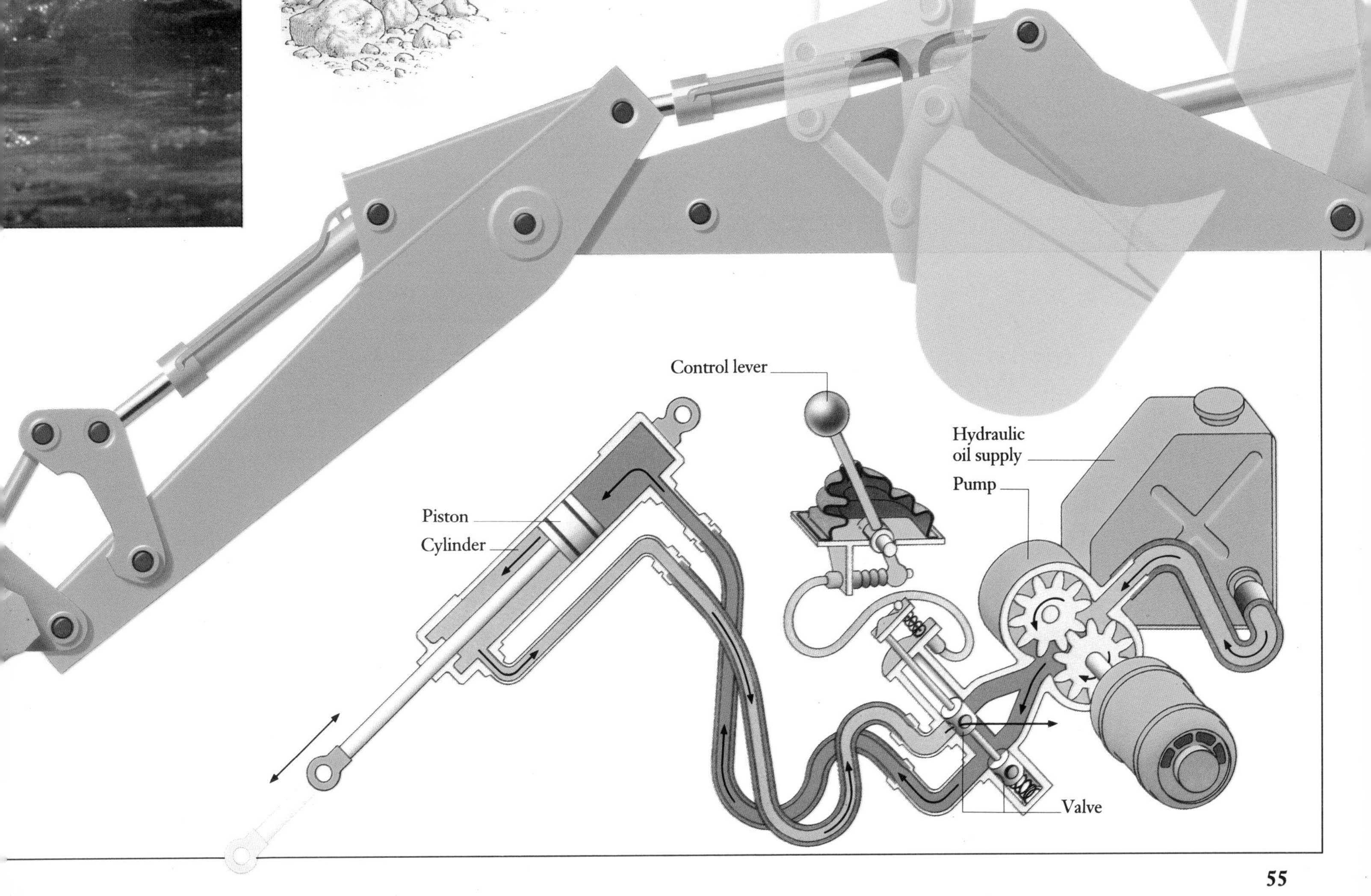

Control lever

Hydraulic oil supply

Pump

Piston

Cylinder

Valve

THE SOLID STATE

Most solids exist in the form of crystals. Their shape reflects the regular arrangement of the atoms or ions within them, which are held together by strong interatomic forces, giving solids the properties of hardness, strength, rigidity and high melting point. A weaker force holds together the molecules in amorphous solids such as glass, whose atoms do not form a regular pattern and more closely resemble those in a liquid. Softer amorphous solids, such as waxes and many plastics, are composed of large molecules held together by weaker intermolecular forces. They lack strength and melt at low temperatures.

Even within the rigid lattice of a crystal, the individual atoms vibrate slightly. The amount of vibration depends on temperature. As a solid is heated, its atoms vibrate more vigorously, taking up more room; this is why most solids expand on heating. At a high enough temperature, the atomic vibrations overcome the interatomic forces and the solid melts to form a liquid.

The hardness of a solid can also be explained in terms of its atomic structure. A good example is the element carbon, which occurs naturally in several forms, or allotropes. The crystalline form of carbon is diamond, in which each carbon atom is chemically bonded to four others in a tight lattice. Diamond is the hardest natural substance known, and it is extremely difficult to cut. It is used in industry in drilling and grinding even the hardest of metals. But in graphite, another allotrope of carbon, each atom is bonded to three others to form layers or sheets that are separated from each other by weaker molecular bonds. As a result, the sheets of atoms easily slide over each other, and graphite is so soft that it is commonly used as a lubricant.

On a scale of hardness devised in 1822 by the German mineralogist Friedrich Mohs, diamond is the hardest, assigned the number 10. The softest is the mineral talc (number 1). Hard metals – such as cast iron – tend to be brittle, and shatter easily. Soft metals, such as aluminum, copper, gold and lead, can be drawn through dies to make wires, or beaten into thin sheets. Gold can be beaten into sheets so thin that strong light passes through them.

When a solid is stretched, its atoms are pulled slightly apart. The interatomic forces try to pull the atoms back to their original positions, and when the stretching force is removed, the solid snaps back to its original dimension. This is called elasticity. Hooke's law, formulated by the English physicist Robert Hooke, expresses the relationship between strain and stress on a solid.

A solid obeys Hooke's law up to a particular stretching force called the elastic limit. Stressed beyond this limit, the solid remains stretched slightly and does not return to its original dimension. Stressed even further, it reaches its yield point, beyond which it continues to stretch with only a slight increase in the stretching force until eventually it breaks. Engineers use machines to stretch materials in this way and measure their tensile strength. Such measurements are essential in the design of aircraft, bridges and ships.

▲ Most solids – even rock and heavy metals – melt if they are made hot enough. Ice melts at 0°C, sodium at 98°C, and caffeine at 238°C. No matter how high or low the melting point, a solid melts when its component atoms or molecules are given enough heat energy for their vibrations to overcome the interatomic forces holding them together.

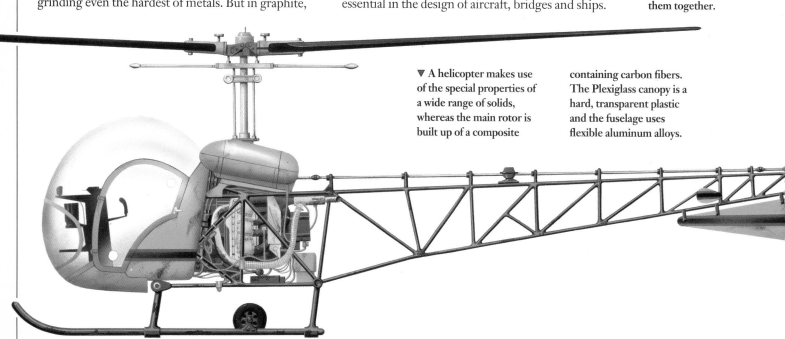

▼ A helicopter makes use of the special properties of a wide range of solids, whereas the main rotor is built up of a composite containing carbon fibers. The Plexiglass canopy is a hard, transparent plastic and the fuselage uses flexible aluminum alloys.

▶ The Eiffel Tower, in Paris, built in 1889, bears testimony to the strength and permanence of metals. It rises to a height of 320 meters and is made of wrought iron. Eight years earlier, Gustave Eiffel had built the iron framework of the Statue of Liberty.

2
FORCE
& Energy

AN OBJECT THAT IS MOVING ALONG – such as a ball rolling along the ground – will go on moving forever unless something stops it. That "something" is a force; in the case of the ball it is friction with the ground, or a reactive force if the ball collides with another object. A force is also needed to start the ball rolling in the first place. The more massive an object is, the greater is the force needed to move it. It also takes a greater force to make an object accelerate quickly.

When a force causes an object to move, energy is used and work is done. The amount of work – measured in joules – is equal to the product of the force and the distance it moves. Energy can therefore be described as the capacity to do work; it also takes various forms. Mechanical energy may exist as one of two kinds of energy: kinetic energy or potential energy. An object, such as a rolling ball, has kinetic energy when it is moving. A weight that releases energy when it falls has potential energy due to its position. A compressed spring also stores potential energy.

Power is the rate of doing work. It takes a certain amount of power for somebody to carry a heavy box up a flight of stairs. But it takes much more power to run up the stairs carrying the same box. Power is measured in watts.

An ocean wave crashes onto the beach, after traveling for hundreds or perhaps thousands of kilometers. In its final onslaught the mass of water contains enough energy to smash rocks and concrete harbors. In the open sea the main water movement is up and down, and this motion can be harnessed to generate electricity.

MATTER ON THE MOVE

WHEN a horse carrying a rider stops suddenly, the rider is thrown forward, often with painful consequences. This is an example of the first law of motion, stated by the English mathematician and philosopher Isaac Newton (1642–1727): an object remains at a state of rest or continues in uniform motion in a straight line, unless it is acted upon by external forces. The object's resistance to a change in its motion results from its mass or inertia; inertia throws the rider forward when the horse stops. A heavyweight rider is thrown harder because of greater momentum, which increases with mass and velocity (speed in a certain direction).

Newton's first law of motion encapsulated the idea that any form of movement must involve at least one force. The second law of motion picked up on his idea of momentum, and states that the rate of a moving object's change of momentum is proportional to (and in the same direction as) the force producing the change. In most cases, the mass of the object does not change, and the law can be simplified to state simply: force equals the product of mass and acceleration.

The third law of motion predicts what happens when two objects are involved. It states that if one object exerts a force on another, there is an equal and opposite force – called a reaction – that the second object exerts on the first. When the gases burn inside a rocket motor, they expand and push equally in all directions. The gases that exert a force on the closed front end of the combustion chamber cause a reaction – a force that pushes in the opposite direction and produces the thrust that propels the rocket.

Unlike a jet engine, a rocket is not propelled by exhaust gases pushing backwards against the air; if this were so, a rocket would not be able to work in outer space, where there is no air. Because its design is based on the application of Newton's third law, a rocket may be technically described as a reaction motor.

The recoil of a rifle illustrates a related principle which results directly from Newton's second and third laws of motion. According to the principle of conservation of momentum, the total momentum of two colliding objects after impact is equal to their total momentum before impact (as long as no external forces come into play). When a

■ A football game provides many examples of Newton's first law of motion: force is needed to move an object or change its direction of movement. The force is the product of an object's mass and acceleration, so the heavier the player, and the faster he accelerates, the more force he brings to bear.

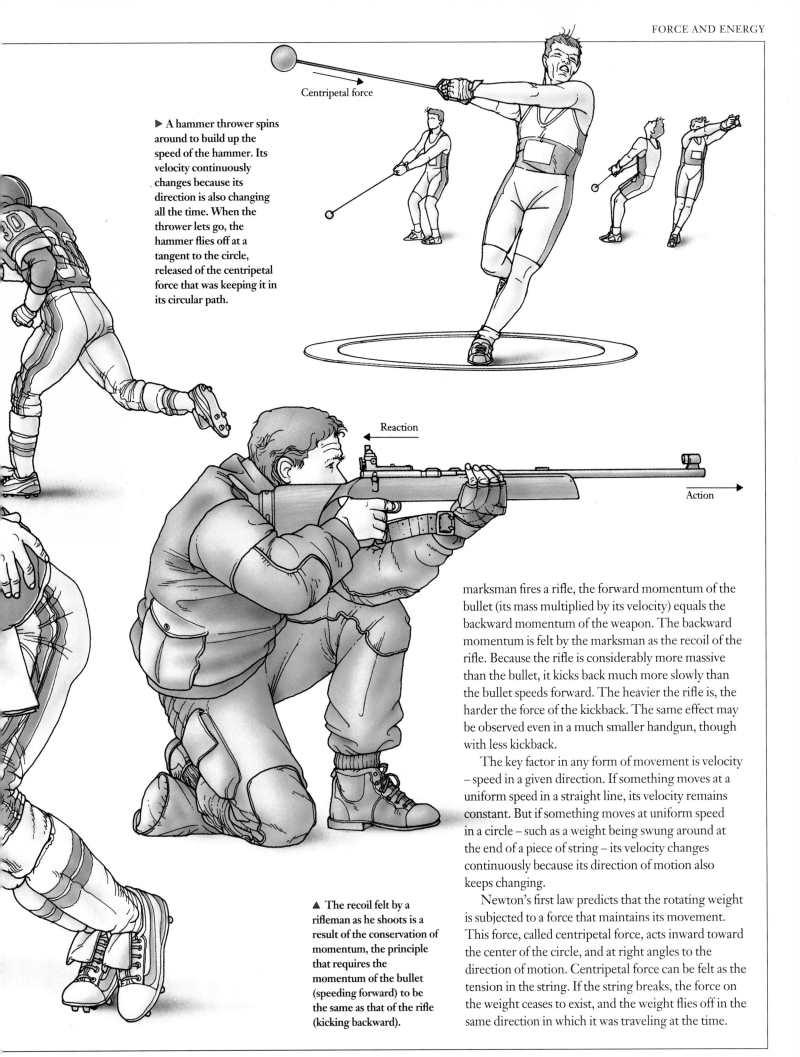

Centripetal force

▶ A hammer thrower spins around to build up the speed of the hammer. Its velocity continuously changes because its direction is also changing all the time. When the thrower lets go, the hammer flies off at a tangent to the circle, released of the centripetal force that was keeping it in its circular path.

Reaction

Action

▲ The recoil felt by a rifleman as he shoots is a result of the conservation of momentum, the principle that requires the momentum of the bullet (speeding forward) to be the same as that of the rifle (kicking backward).

marksman fires a rifle, the forward momentum of the bullet (its mass multiplied by its velocity) equals the backward momentum of the weapon. The backward momentum is felt by the marksman as the recoil of the rifle. Because the rifle is considerably more massive than the bullet, it kicks back much more slowly than the bullet speeds forward. The heavier the rifle is, the harder the force of the kickback. The same effect may be observed even in a much smaller handgun, though with less kickback.

The key factor in any form of movement is velocity – speed in a given direction. If something moves at a uniform speed in a straight line, its velocity remains constant. But if something moves at uniform speed in a circle – such as a weight being swung around at the end of a piece of string – its velocity changes continuously because its direction of motion also keeps changing.

Newton's first law predicts that the rotating weight is subjected to a force that maintains its movement. This force, called centripetal force, acts inward toward the center of the circle, and at right angles to the direction of motion. Centripetal force can be felt as the tension in the string. If the string breaks, the force on the weight ceases to exist, and the weight flies off in the same direction in which it was traveling at the time.

FORCE OF GRAVITY

G RAVITY is an important phenomenon that applies in physics, astronomy, space science, construction and engineering. It is a force of attraction that acts between objects because of their masses.

Mass is a measure of the quantity of matter. An object has the same mass on Earth, on the surface of the Moon or in outer space, because it contains the same amount of matter wherever it is. But objects on Earth also have weight, which is the force of Earth's gravity acting on them. Weight is most properly measured in newtons, although for convenience it is often expressed in mass units such as kilograms and pounds.

Weight can be stated as mass multiplied by acceleration. For an object on Earth, therefore, weight is the object's mass multiplied by the acceleration due to gravity (also called the acceleration of free fall), a constant quantity in physics equal to nearly 9 meters per second per second (9 m/sec^2). The acceleration due to gravity on the Moon, however, is only about 1.6 m/sec^2, which is why objects on the Moon weigh only about one-sixth as much as they do on Earth.

Gravity is a force that can act at a distance. Indeed, the gravitational attraction of the Moon – although fairly weak – is strong enough over a distance of 382,000 kilometers to affect sea levels on Earth and cause the tides. With increasing distance, gravity becomes weaker. On Earth, the force of gravity acts between the center of the Earth and the center of the object.

▶ The classic experiment performed by the British physicist Henry Cavendish in 1798 provided the first measurement of the gravitational constant G and, derived from this, the mass of the Earth. He used an apparatus called a torsion balance to measure the gravitational force of attraction between a pair of massive lead spheres and two much smaller, lighter spheres. The attraction of the lighter spheres to the heavier twisted the wire suspension, and the twisting force, or torsion, could be measured.

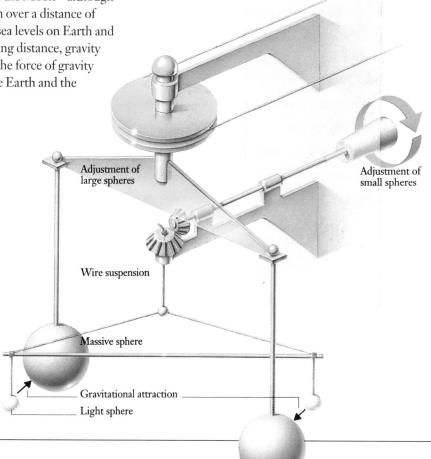

Adjustment of large spheres

Adjustment of small spheres

Wire suspension

Massive sphere

Gravitational attraction

Light sphere

■ For the Space Shuttle to go into orbit, the craft must be given enough speed to escape totally from the Earth's gravity. This speed, called the escape velocity, equals 11.2 kilometers per second. Once in orbit FAR LEFT, the astronauts are in free fall (moving with the acceleration due to gravity), and are "weightless".

▼ A satellite can orbit at any altitude as long as it is traveling fast enough to overcome the Earth's gravity (and remain in free fall). A satellite just above the limits of the Earth's atmosphere (at an altitude of about 230 km) orbits

The gravitational force of attraction between two objects is proportional to the product of their masses and inversely proportional to the square of the distance between them. This relationship, called the universal law of gravitation, was worked out in about 1666 by the English scientist Isaac Newton (1642–1727). The gravitational force between two objects acts between their centers of mass – the place at which all their mass appears to be concentrated. Sometimes also called the center of gravity, the center of mass of an object has an important influence on its stability.

The stability of an object can be defined by imagining a straight line drawn vertically down through its center of mass. With a pyramid standing on its base. A line drawn down from its center of mass passes through the base; the pyramid is said to be in stable equilibrium. If it is tipped slightly, its weight acts downward to pull it back into a stable position. But if the pyramid is balanced upside down on its point, the slightest movement makes it topple over; it is in unstable equilibrium.

A sphere, or cylinder lying on its side, is in neutral equilibrium – if it is displaced sideways, the center of mass still acts downwards through the point of contact. Objects in stable or neutral equilibrium stay where they are unless acted on by an external force.

once in less than two hours. At an altitude of about 36,000 km, the period of orbit is 24 hours; the satellite orbits once as the

Earth spins once on its axis, and so it appears to remain stationary overhead: this phenomenon is called a geostationary orbit.

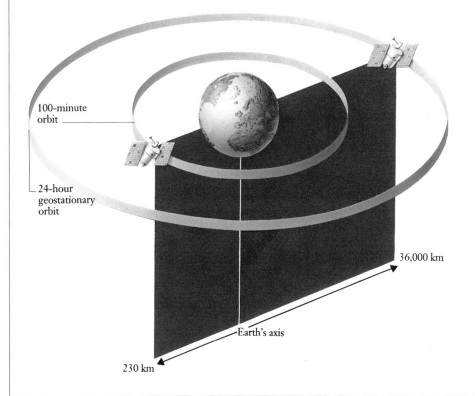

100-minute orbit

24-hour geostationary orbit

36,000 km

Earth's axis

230 km

MECHANICAL ENERGY

A<small>N OBJECT</small> that has mass and is moving possesses
kinetic energy, equal to the product of half
its mass and the square of its velocity. It may be
translational kinetic energy, for an object moving in a
line, or rotational kinetic energy, for something
spinning on its axis. Translational kinetic energy is
proportional to the square of an object's velocity. This
fact has important consequences. For example, if an
object's velocity doubles, its kinetic energy increases
four times. This is why the
speed of a road vehicle –
rather than its mass – is the
chief factor that results in
damage in a collision. A car
traveling at 135 km/h (80
mph) has 16 times the
kinetic energy of – and
therefore a much greater
impact than – a similar car
that is traveling at a speed of only 32 km/h (20 mph).

KEYWORDS

ALTERNATING CURRENT
KINETIC ENERGY
MASS
POTENTIAL ENERGY
SIMPLE HARMONIC
    MOTION

An object with mass can also possess energy
because of its position – called gravitational potential
energy – or because it is deformed, like a stretched or
compressed spring – called elastic potential energy.
The value of gravitational potential energy possessed
by or "stored" in an object is equal to the product of its
mass, its height and the acceleration due to gravity (the
acceleration of free fall). Thus, the heaver it is, or the
higher it is, the greater is the potential energy.

These are all various forms of mechanical energy,
and they can all be made to do work. The translational
kinetic energy of a rapidly moving pool cue, for
example, transfers to the ball and makes it roll away;
the rotational kinetic energy stored in a heavy flywheel
can be made to operate machinery. Gravitational potential energy is not usually apparent until it is converted into kinetic energy. The gradually falling weights
of an old-fashioned pendulum clock drive around the
cog wheels, and the water stored behind a dam can be
made to release its potential energy to turn the blades
of a turbine. A simple example of stored elastic potential energy is a drawn bow, which releases its energy in
an instant to fire an arrow to its target.

In some systems potential and kinetic energy are
continuously interchanged. One example is a pendulum, which consists of a weight swinging at the end of
a rod or string. At the top of a swing, the weight has
only potential energy. Then as it swings, gradually
losing height, this energy is converted to kinetic energy; at the lowest point of the swing it has only kinetic
energy and no potential energy. The potential energy
is gradually restored as the weight rises on the second
half of its swing, at the end of which it is momentarily

stationary and has no kinetic energy. The rising weight does work against the force of gravity.

A swinging pendulum also illustrates periodic, or oscillating, motion – motion that varies predictably with time. If the sideways displacement of the pendulum's weight is plotted against time, the resulting shape is known as a wavy sine curve. A similar curve is obtained by plotting against time the displacement of a weight oscillating at the end of a vertical spring.

The time for one complete oscillation is called the period of the motion, and the maximum displacement from the equilibrium position – sideways for the pendulum and up or down for the oscillating spring – is known as the amplitude. Any motion that produces a sine curve in this way is called simple harmonic motion (SHM). There are many other examples in physics, such as the rapid voltage oscillations of an alternating electric current or radio wave.

Even in oscillating systems, the sum of the kinetic and potential energies always stays the same. This is one example of a wider – and important – principle in physics, called the conservation of energy. Formally, it states that the total amount of energy in any system remains constant, even though changes of energy from one form to another may take place.

◀ Potential energy is stored energy. The energy stored in the mass of water held back by a dam is released, and converted to kinetic energy, when it flows to turn the blades of a turbine.

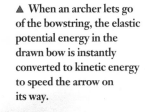

▲ When an archer lets go of the bowstring, the elastic potential energy in the drawn bow is instantly converted to kinetic energy to speed the arrow on its way.

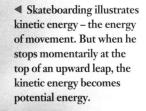

◀ Skateboarding illustrates kinetic energy – the energy of movement. But when he stops momentarily at the top of an upward leap, the kinetic energy becomes potential energy.

SIMPLE MACHINES

THE Greek mathematician Archimedes (c. 287– c. 212 BC) is credited with saying that, given a lever long enough, he could move the world. The lever was one of the simplest of early machines. Using a long pole pivoted over a flat stone, a person can lift and move a massive object such as a bolder. The same mechanical principles that apply to early machines such as levers and pulleys are still behind the designs of the most complicated modern machines, from gear boxes to escalators and giant cranes. All such mechanical devices multiply the limited force available to humans and animals to help them make things move.

The simplest kind of lever, described above, has the pivot situated between the point at which the effort is applied and the point at which the load is raised. It is called a class one lever. Another example is a seesaw, on which a small child sitting at one end can lift a large adult sitting nearer the pivot on the other side. This example illustrates the key reason for using levers: if the pivot is closer to the load than the effort, a small effort can produce a large effect. A long lever that uses a small effort to lift a heavy load has a large force ratio or mechanical advantage, defined as the load (the output force) divided by the effort (the input force). Class one levers also combine in pairs to form scissors and shears.

There are two other types of levers. A class two lever has the effort and the load on the same side of the pivot (known technically as a fulcrum), as in a wheel-barrow. In a class three lever, the effort is between the load and the fulcrum, as in a pair of tweezers or the muscles that bend a human arm.

For a long lever, the effort has to move a long way to produce only a small movement of the load. It is described as having a small distance ratio (or velocity ratio) – the distance moved by the load divided by the distance moved by the effort. The efficiency of any machine is the usable energy output divided by the energy input. For a lever, it is the same as the force ratio divided by the distance ratio. Overall, the class one lever is an efficient machine.

Pulleys and gears also provide a mechanical advantage, allowing a small effort to move a large load. A single pulley merely changes the direction of the effort and, because of the friction in the pulley, may actually need an effort that is larger than the

■ There are two basic types of simple machines: the lever and the inclined plane. Examples of the three classes of levers are illustrated here. The wheel and axle is an extension of the lever principle. It is employed in a windlass, gear wheels, belt drives and pulleys. An inclined plane – which provides mechanical advantage by moving a load up a slope instead of lifting it vertically – is the basis of the screw jack as well as ordinary nuts and bolts.

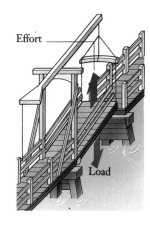

▶ A simple lift bridge can be built using a class three lever. As in class two, the load and effort are both on the same side of the fulcrum, but here the effort is nearer the fulcrum than the load. A considerable effort is needed to lift the load, and as the effort is released, the load returns to its former position. This can be an obvious advantage with a canal bridge, which should "fail safe" when in the down position.

◀ The lifting section of a wine-bottle opener is achieved by a pair of class one levers. The load (the corkscrew) is on one side, and close to, the fulcrum (the pivot). The effort, exerted by the hand, is on the other side of the fulcrum and much further from it. As a result, a comparatively small effort exerts a very large force to lift the cork from the bottle.

load. But for two pulleys (neglecting friction), the force ratio is 2 – a given weight can lift a load weighing twice as much. For four pulleys the force ratio is 4, and so on: the mechanical advantage equals the number of pulleys.

Distance ratios get smaller as the number of pulleys increases, because the effort has to move longer distances to raise the load. The theoretical efficiency remains the same (at 100 percent), but in practice, friction takes a greater and greater toll, so that multiple pulleys are much less efficient than single ones. But if a pulley block is the only machine available – as it was for centuries – providing more force to overcome friction still enables muscle power to be multiplied and to achieve results impossible for the unaided human.

An inclined plane hardly seems to be a machine at all. At its most basic, it is a slope up which a load must be pushed. But it is much easier to push the load up the slope than to lift it vertically to the same overall height. There is a mechanical advantage, equal to the length of the slope divided by the height – the gentler the slope, the greater the mechanical advantage.

Two applications of the inclined plane are the wedge and the screw. A wedge may be driven under a load to lift it, or into something to split it. If an inclined plane is wrapped around a cylinder, a helical inclined plane results: a screw. Turning the screw forces it into a material, like a spiral edge.

▶ A sack barrow is an example of a class two lever in use. Here, both the load and the effort are on the same side as the fulcrum (the axle of the wheels). But because the load is nearer to the fulcrum than to the effort, a person exerting maximum human effort can lift – and move – a load in excess of his or her own body weight. Lighter loads require much less effort.

Effort

Load

HEAT ENERGY

WHEN an object gets hot, the heat energy is stored in its atoms, which continuously vibrate – the more they vibrate, the hotter the object becomes. Heat is a form of kinetic energy, the energy of motion of the vibrating atoms. If an object is made cold enough, its atoms will stop vibrating altogether. The temperature at which this happens, called absolute zero, has never been reached, though scientists in July 1993 achieved a record low temperature of .000007 degrees Kelvin (within a hundred-thousandth of a degree of absolute zero).

Heat is regarded as a separate type of energy, capable of being converted into all other forms. If anything is made hot enough it gives off light, and in a thermocouple – a type of electrical circuit – heat is converted directly into electricity. Other forms of energy can also be turned into heat: electricity flowing in a high-resistance wire, and friction between two moving surfaces in contact, both generate heat.

Heat can travel from place to place. In a bar of metal heated at one end, for example, the vibration of the hot atoms is passed on to their neighbors so that heat gradually travels along the bar toward the unheated end. This type of heat transfer is called conduction, and materials such as metals are good conductors of heat.

Heat can also be transferred in the motion of a hot gas or liquid. A warm gas is less dense than a cold one and so it tends to rise, causing air currents in a heated room, or large movements of air which result in changes in patterns of weather. This type of heat transport is called convection, and the resulting movements of gas or liquid are convection currents.

Heat can also move by radiation – the heat from the Sun reaches the Earth through the vacuum of space by this method. Any object whose temperature is above absolute zero emits heat radiation, although the rates of emission become significant only at high temperatures. The amount of heat emitted depends also on the nature of the surface, a matte black surface being a much better radiator than a shiny silver one.

Another way of defining heat is to regard it as energy in transit from one object to another because they have different temperatures. A knowledge of how heat travels can be useful in stopping unwanted heat flow. One way of keeping something warm is to surround it by a poor conductor – a thermal insulator such as plastic foam – to prevent heat from being

KEYWORDS

ABSOLUTE ZERO
CONDUCTION
CONVECTION
HEAT
RADIATION
THERMOCOUPLE

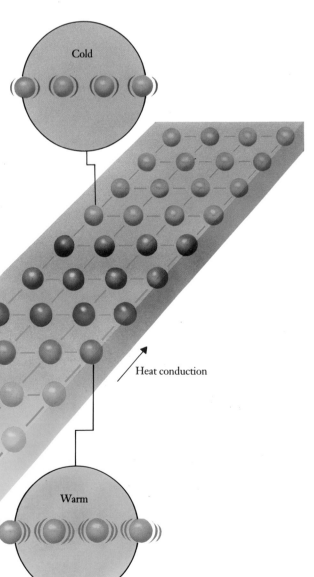

Cold

Hot

Warm

Heat conduction

◄ A metal bar heated at one end illustrates the kinetic theory of heat – that is, the atoms in any object at a temperature above absolute zero are in a state of constant vibration. The hotter they get, the more they vibrate; and the vibrations are passed on to neighboring atoms. This last effect accounts for the conduction of heat along solid objects. It also explains why objects expand when heated – the atoms take up more room. The increase in thermal vibration and hence the extent of linear expansion depend directly on the temperature.

conducted away. In a similar way, quilted or padded clothing helps keep people warm in cold climates. A vacuum flask also prevents the other kinds of heat transfer, keeping cold liquids cold or hot liquids hot. The vacuum between the thin walls of the container prevents heat moving by convection (there is no air to convect and carry the heat), and the silver mirror on the walls of the vessel minimizes transfer by radiation.

Heat is measured in joules. But some applications use different, and older, heat units. For example, dieticians express the energy content of food in kilocalories (usually expressed as Calories, with a capital C). A kilocalorie is the amount of heat needed to raise the temperature of a kilogram of water through 1 degree Celsius. The calorific value of a food is the amount of energy released when it is completely burned. Sugar has 39 calories per gram; fats (such as butter) have 76 calories per gram.

▲ Many solids, including metals, soften on heating – a phenomenon used by a blacksmith when beating hot iron to make a horseshoe. Heat also affects many other physical properties of materials, such as their ability to conduct electricity.

▷ Many soaring birds make use of thermals to climb effortlessly higher in the sky without flapping their wings. A thermal, in turn, results from the convection of air. As the Sun warms the ground or the water at the surface of the sea, the air near the surface becomes warm. Warm air is less dense than cooler air and so it rises. The upward convection current is called a thermal.

Bird's flight path

Thermal

Heated air

Warm soil

HEATING AND COOLING

MATERIALS differ in their ability to hold heat. This is measured as their specific heat capacity – the amount of heat needed to raise one kilogram of a substance through 1 degree Kelvin.

When a pure solid is heated, its temperature rises in a regular way until, when it reaches a certain temperature, it continues to absorb heat without its temperature rising further; the substance then melts and its temperature starts rising again. The extra heat needed to make this happen is called the latent heat of fusion. The same amount of heat is released when a liquid is cooled until it freezes. More latent heat – the latent heat of vaporization – is also required to make a liquid at its boiling point change into a gas or vapor. These latent heats weaken the attractive forces between atoms or molecules, turning a solid into a liquid or a liquid into a gas. When a substance cools and turns from gas to liquid, or liquid to solid, the latent heats are emitted.

As a hot object cools, it transfers heat to its surroundings. To make it colder than the surroundings, heat has to be removed from it. When a gas flows through a small hole into a larger container, its temperature drops as its pressure falls (because heat is used up in pushing the gas molecules farther apart). Called the Joule–Kelvin, or Joule–Thompson, effect, this is the principle of a refrigerator, in which the heat absorbed by an expanding gas is taken from the surrounding interior of the refrigerator. The effect also explains why an aerosol spray feels so cold as it emerges from its container, though the exterior of the container does not feel cold itself.

The Joule–Kelvin effect can be used to liquefy a gas and create very low temperatures. Usually the gas, such as air, is first compressed and cooled to about -25°C in a refrigerator, and then cooled even further to -160°C by making it do work by expansion in a turbine. Finally the cold air expands through a narrow aperture and turns to a liquid (at about -180°C). Other gases liquefy at much lower temperatures, such as hydrogen (-259°C) and helium (-271°C).

Such is the interest in the properties of materials at extremely low temperatures that its has given rise to a new discipline of science known as cryogenics, which also encompasses the techniques for producing such cold environments. The usual way of rapidly cooling substances under study is to immerse them in a bath of liquid gas, which can reach temperatures as low as 3 K (-270°C). Researchers in laboratories have achieved temperatures that come within a millionth of a degree of absolute zero – a purely theoretical temperature that is actually impossible to achieve.

Materials can have very strange properties at such low temperatures. At a low enough temperature, liquid helium , for instance, loses all its viscosity and becomes a superfluid. It climbs up the walls of its container, leaving it empty.

Some metals lose all their electrical resistance and become superconductors. If an electric current is made to flow in a circuit of superconductors, it goes on flowing forever without the need for an external source of voltage. Superconducting magnets have various industrial applications, in devices such as nuclear magnetic resonance machines and magnetic levitation trains.

A search continues for materials that will superconduct at higher temperatures, and in early 1994 it was reported that a superconducting material had been found that would operate at temperatures as high as -25°C, which would open the way to more widespread industrial applications.

▶ **A domestic refrigerator, like all other types of refrigerator, works by removing heat from its contents. The heat is absorbed by an ice-cold liquid – a refrigerant – circulating in narrow pipes. The liquid becomes warmer and turns to a gas, which is forced through a narrow jet (expansion valve). This process rapidly cools the gas, which is then compressed to liquefy before it is recirculated. Heat is evolved during the compression of the refrigerant and dissipated to the outside air.**

A similar principle is used as a space heater in the device known as the heat pump. This effectively reverses the process of the refrigerator. It pumps heat from the exterior into the interior space.

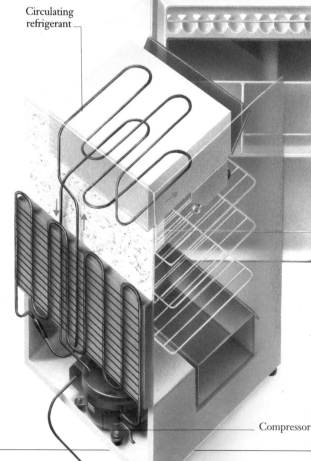

Circulating refrigerant

Compressor

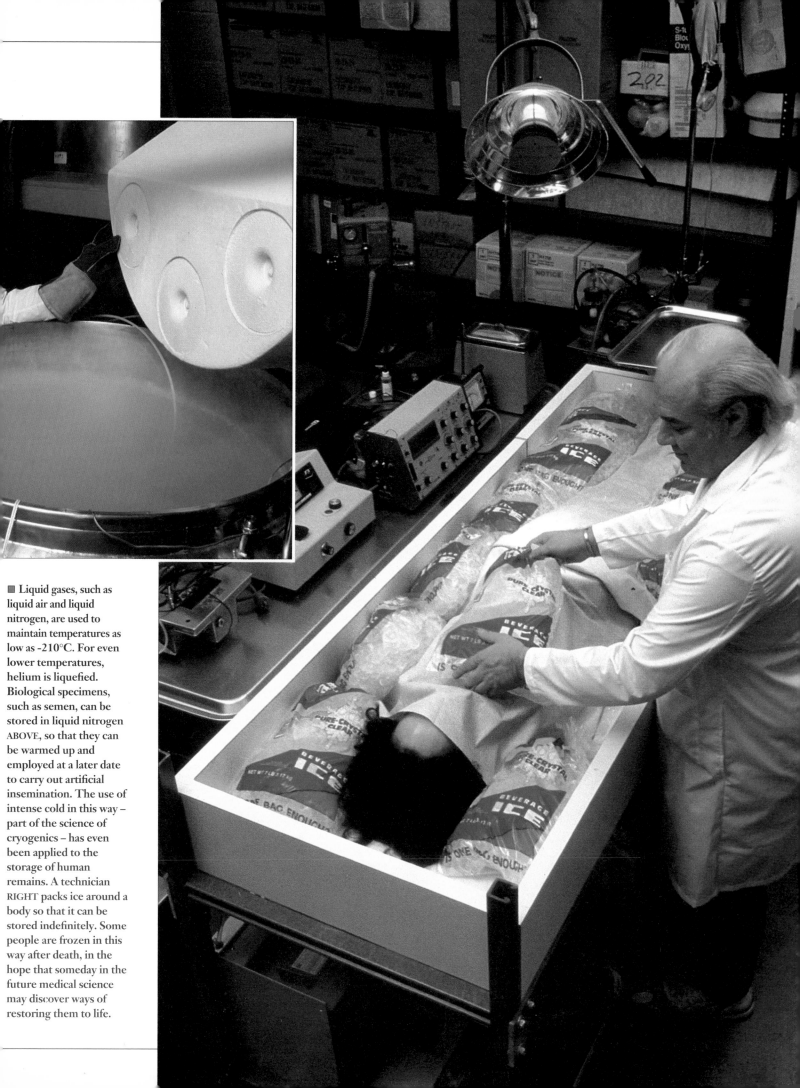

Liquid gases, such as liquid air and liquid nitrogen, are used to maintain temperatures as low as -210°C. For even lower temperatures, helium is liquefied. Biological specimens, such as semen, can be stored in liquid nitrogen ABOVE, so that they can be warmed up and employed at a later date to carry out artificial insemination. The use of intense cold in this way – part of the science of cryogenics – has even been applied to the storage of human remains. A technician RIGHT packs ice around a body so that it can be stored indefinitely. Some people are frozen in this way after death, in the hope that someday in the future medical science may discover ways of restoring them to life.

MEASURING AND USING HEAT

TEMPERATURE is the degree of hotness of an object. It can be measured in terms of any of several physical properties that change with temperature. An ordinary thermometer makes use of the expansion of a liquid in a narrow glass tube. The expansion of the liquid – usually dyed alcohol or mercury – is shown on a scale calibrated in degrees. The positions of the calibrations are determined by two so-called fixed points, which for an ordinary thermometer are usually the freezing and boiling points of water.

On the Celsius (formerly centigrade) scale, these temperatures are 100 degrees apart. On the Fahrenheit scale they are 180 degrees apart (freezing point of water 32°F, boiling point 212°F). Physicists and other scientists often use the absolute, or Kelvin, temperature scale, which runs from absolute zero (0 K) and has degrees that are the same size as Celsius degrees. On this scale, the freezing point of water is just over 273 K and its boiling point is 373 K.

The electrical resistance of a metal – its ability to carry an electric current – also varies with temperature. This phenomenon is used in a platinum resistance thermometer, which measures temperature by the resistance of a piece of platinum wire. Very high temperatures can be measured by making use of the fact that the speed of sound through a gas depends on its temperature. A microphone on one side of the furnace picks up a series of clicks generated by an electric spark on the other side. A computer measures the temperature by calculating the time taken by the sounds to reach the microphone.

KEYWORDS

CARNOT CYCLE
HEAT ENGINE
INTERNAL COMBUSTION ENGINE
TEMPERATURE
THERMODYNAMICS

▶ Glass and modern ceramics are versatile materials that can be used to make a wide range of heat-resisting objects. They are all formed at high temperatures in furnaces such as the one shown here, which have to be carefully controlled and monitored to ensure the quality of the product. An instrument for measuring such high temperatures is called a pyrometer, and the usual type compares the color of the hot object with an electrical standard.

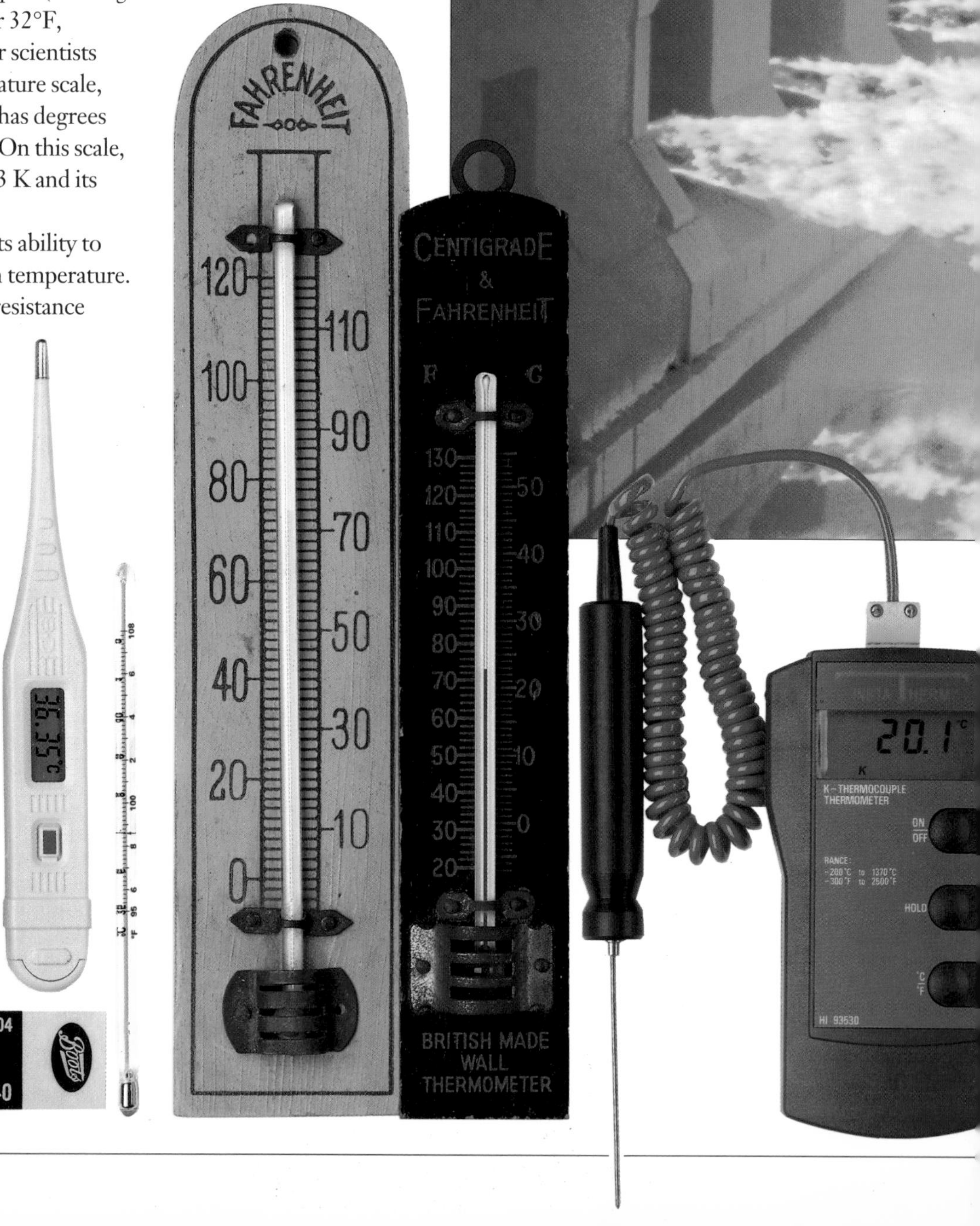

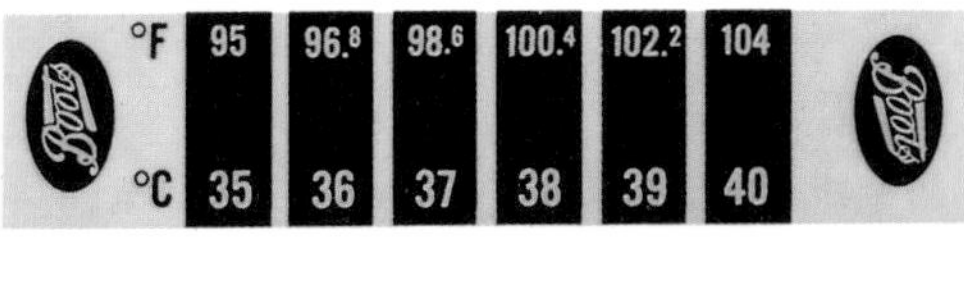

▶ Thermometers range from liquid-crystal devices and the liquid-in-glass type to electronic thermometers that use a thermoelectric probe to produce a digital display of temperature.

°F	95	96.8	98.6	100.4	102.2	104
°C	35	36	37	38	39	40

To put heat to good use, it is often changed into another form of energy. The function of a heat engine – such as a gasoline engine, diesel engine, gas turbine or rocket – is to convert heat into mechanical energy. In each of these, heat makes gases that are the products of combustion expand and move a piston or turbine, or create thrust. The efficiency of such engines depends mainly on the difference in temperature (and therefore the energy content) between the gases before and after they have done their work. This way of studying engines was first suggested by the French engineer Nicolas Sadi Carnot (1796–1832).

Heat engines work on a cycle of changes involving the intake, combustion, expansion and exhaust of hot gases. But even a theoretically perfect engine is less than 100 percent efficient. In real engines, efficiency – the ratio of heat input to heat output – is usually between 40 and 50 percent because of energy losses such as friction, waste heat and sound. In these engines, heat flows from a hot region to a cooler one, because heat will not flow from a colder object to a hotter one unless external energy is supplied – a principle known as the second law of thermodynamics. The first law of thermodynamics reflects the principle of conservation of energy: in a closed system, the total amount of energy (such as heat) is constant.

Thermodynamics also involves another quantity, known as entropy. For a system possessing energy, entropy is a measure of the unavailability of that energy to do useful work – that is, it is a measure of the system's disorder. If entropy increases, less energy is available for work . But, like heat energy, entropy decreases with temperature, and the third law of thermodynamics states that, at absolute zero, the entropy of a perfect crystal is zero.

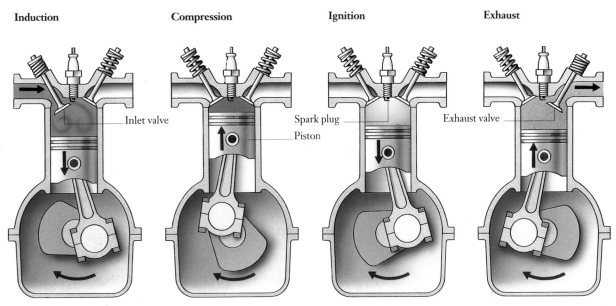

▶ One of the most important applications of heat is in a heat engine, particularly in an internal-combustion engine. The first of these to be developed commercially – gas and, later, oil engines – made use of the four-stroke (or Otto) cycle. The key stages of this cycle are induction (the fuel/air mixture enters the cylinder), compression (the piston moves up), ignition (the spark plug explodes the fuel and the piston moves down), and exhaust (spent gases leave the cylinder).

Induction   **Compression**   **Ignition**   **Exhaust**

Inlet valve   Spark plug   Exhaust valve
Piston

ELECTRICITY
& *Magnetism*

3

A S EARLY AS THE SECOND CENTURY BC, magnets were used by Chinese navigators. These ancient sailors found that if they suspended a piece of the mineral lodestone on a thread, it always lined up in a north-south direction and was used as a primitive compass. The ancient Greeks discovered that a piece of amber rubbed with fur attracted pieces of straw and other scraps, in much the same way that a magnet attracts small pieces of metal. However, in this case the origin of the attraction is not magnetism but static electricity. The Greek word for amber is elektron, which is the root of the words electron and electricity. Electricity and magnetism are produced by the behavior of electrons and other particles in atoms.

The discovery of the intimate relationship between magnetism and electricity had to wait for the work of 18th- and 19th-century physicists, who established that an electric current flowing in a wire produces a magnetic field, and that a magnetic field can be made to generate a current in a wire. These phenomena, called electro-magnetism, are the basis of most of the electrical machines we have today. Electromagnets, electric motors and generators, transformers, microphones, loudspeakers and even electric bells and buzzers all make use of the interplay between electricity and magnets.

The awesome power of static electricity is demonstrated when a flash of lightning strikes a lightning conductor at the top of one of the 411-meter towers of the World Trade Center in New York City. During a lightning strike, the sudden discharge of an electric potential of a million volts per meter ionizes the air and allows it to conduct huge currents in excess of 10,000 amps.

MAGNETS AND FIELDS

A MAGNET is piece of metal that attracts or repels a similar nearby piece of metal. The effect can be traced to the subatomic particles that make up the atoms of the metal.

As electrons (which are negatively charged) orbit the nuclei of the atoms, they spin and generate a small magnetic field. The tiny atomic magnets line up with each other to form magnetic regions called domains. In a piece of metal, such as iron or steel, there are millions of domains, some pointing one way, some another, so that there is no overall magnetic field. But if the metal is placed in an external magnetic field, the domains line up parallel with the field and to each other. Their individual tiny fields combine to form a single large one, and the metal becomes a magnet.

Three kinds of magnetic properties depend on the number of electrons and how they spin: ferromagnetic, paramagnetic or diamagnetic. A ferrromagnetic material – such as iron, cobalt or nickel – is one in which, within the atoms in the domains, the spins of the electrons line up in the presence of an external magnetic field. Below a certain temperature the magnetization persists even when the external field is removed, and the material becomes a permanent magnet. Ferrites are ceramic substances (a combination of cobalt, zinc or nickel with an iron oxide) that are ferromagnetic and can be used to make extremely powerful permanent magnets.

Paramagnetic substances acquire magnetic properties in the direction of an external magnetizing field because their component "atomic magnets" line up. But the magnetization disappears when the external field is removed. In a few substances – which are diamagnetic – the temporary magnetization is in the opposite direction to the external field.

In a bar-shaped magnet, the magnetic field appears to originate at a point near one end of the bar and extend in space, curving around to a point near the other end. These points are called the poles of the magnet – north and south – and the field can be represented as lines of force joining them. Magnetic poles always occur in north-south pairs. A line of force can be thought of as the path a single pole would take as it moves in response to the magnetic forces acting on it.

Another property of magnetic poles is that similar ones – such as two north poles – repel each other; dissimilar ones attract each other. Their magnetic

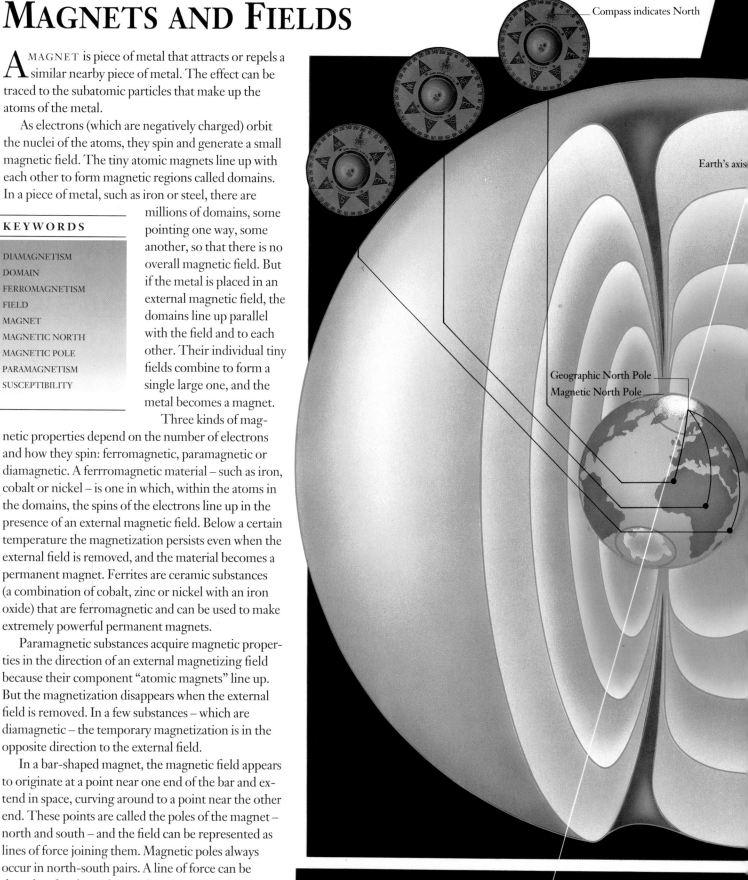

Compass indicates North

Earth's axis

Geographic North Pole
Magnetic North Pole

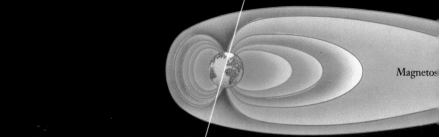

Magnetos

76

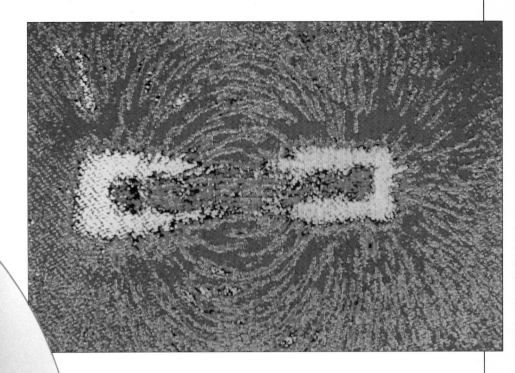

▶ The magnetic field – lines of force – around a bar magnet can be shown by sprinkling iron filings on paper above the magnet.

◀ Electric currents in the molten iron core of the Earth give rise to a magnetic field. The Earth behaves as if there is a huge bar magnet along its axis. A compass has a magnetized needle mounted horizontally, and in the Earth's field its north-seeking pole points to magnetic north. A similar needle mounted vertically is called a dip needle. Near the Equator it is horizontal, but farther north (or south) it dips until, at the poles, it points vertically downwards.

fields either join together or push each other apart. In fact, any two magnetic poles exert a force on each other which is proportional to the product of their strength divided by the square of the distance between them. For this reason, a magnetic field falls off fairly rapidly with distance from the magnet.

The needle in a compass is a small pivoted magnet. Its north-seeking end (actually a north pole) always points in the direction we call north. For this to happen, there must be a south magnetic pole near the Earth's North Pole. It is as if the Earth has a huge bar magnet along its axis, making compass needles all over the world align with the Earth's magnetic field and point north and south. The magnetic poles do not coincide exactly with the geographic poles, however, and they move slowly from year to year. Navigators must take account of this when using a compass.

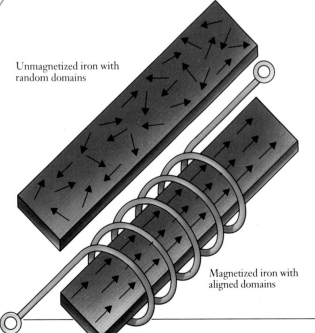

Unmagnetized iron with random domains

Magnetized iron with aligned domains

◀ In an unmagnetized iron bar, the molecular magnets are aligned randomly. When the bar is placed in a coil of wire carrying an electric current, the magnetic field created by the coil aligns the magnets and the iron becomes permanently magnetized. A magnet can also be made by aligning an iron bar with the Earth's magnetic field (by pointing it due north) and stroking it with another magnet, or by striking it with a hammer.

◀ The Earth's magnetic field extends into space for thousands of kilometers as the magnetosphere, which – despite its name – is distorted into a teardrop shape because of the effects of solar wind. Many other planets have similar fields.

STATIC ELECTRICITY

W HEN a nonmetallic object, such as a plastic comb, is rubbed on a piece of dry cloth or fur, it is able to attract light objects such as hair, like a magnet picking up paper clips. This phenomenon is caused by changes in the atoms in its surface. Friction "rubs off" some of the electrons on the atoms of the comb. Because electrons carry a negative charge, the comb is left with a positive charge. Static electricity is the accumulation of these electrostatic charges.

KEYWORDS

CONDUCTOR
ELECTRIC CHARGE
ELECTRIC POTENTIAL
INSULATOR
POTENTIAL DIFFERENCE
STATIC ELECTRICITY

Two combs charged up in this way repel each other. But when a charged comb is brought near scraps of paper, its positive charge causes electrons to accumulate on the closest part of each piece of paper, giving them a negative charge. The negatively charged paper is then attracted to the positively charged comb. The transfer of charge is called electrostatic induction.

A key fact of electric charges is that dissimilar charges attract each other and similar charges repel. A charged object influences the region surrounding it: in other words, a charge creates an electric field. The strength of the attraction or repulsion between the charges depends on the strength of the charges and decreases with distance.

When a nonmetallic material – an insulator – is placed in an electric field, the field exerts a force on the electrons in it. If the field is strong enough, the electrons are torn from their atoms and flow through it, carrying charge. This is what happens in lightning. The huge electric field between charged clouds breaks down the insulating properties of the air between them, or between the clouds and the ground, and a giant spark jumps across. There is an accompanying flash of light and a shock wave in the air. The purpose of a lightning conductor on a tall building is to "capture" the charge and carry it safely to the ground.

▶ During a thunderstorm, objects such as solitary trees or tall buildings take on a positive charge, induced by the huge charge on the base of the thunder cloud. Such objects thus become likely targets for a lightning strike.

▶ Electrostatic paint spraying uses static electricity. The object to be painted (such as a car) is given an electric charge, and the paint droplets leaving spray guns are given an opposite charge, which makes the paint stick.

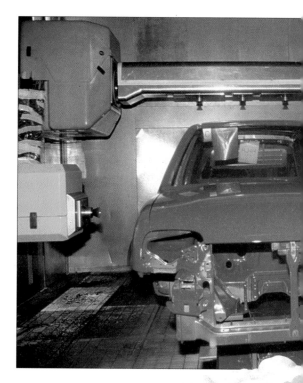

▼ Separation of charges in a thunder cloud BELOW LEFT leads to potentials of a million volts per meter. Eventually the cloud becomes discharged BELOW RIGHT as a huge spark of lightning ionizes the air on a track to the ground. The spark may range in length up to 30 kilometers, and travel at 100 million meters per second – nearly one-third the speed of light.

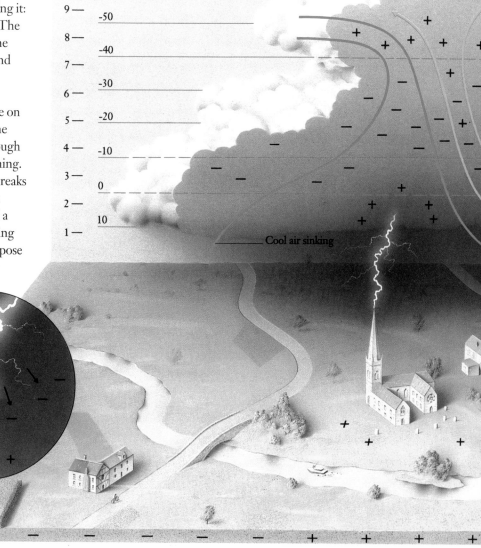

If one object has more charge than another, it is said to have a higher electric potential. When two charged objects are connected, positive charges flow from the one of higher potential to the one with lower potential until their potential is the same. Negative charges move in the opposite direction. The greater the initial potential difference between the two, the more readily charge flows between them. Another term for potential difference is voltage, and such differences are measured in volts. The amount of charge on an object is measured in coulombs.

Charge flow involves changes in energy and therefore in work. Large potential differences are produced by various types of electrostatic machines, such as a Van der Graaff generator, which is used to create high voltages – up to 6 million volts, equivalent to a bolt of lightning – for particle accelerators and other research apparatus. Charges of this magnitude are released in a single spark and accelerated through a series of strong magnetic fields for high-energy experiments.

Groups of atoms, single atoms and even subatomic particles can carry electric charges. Any molecule (group of atoms) or single atom carrying a charge is called an ion. The metals sodium, calcium and iron, for example, form positive ions whereas chlorine and bromine form negative ones. Ions are the current carriers in discharge tubes and fluorescent tubes.

The chief subatomic particles are the electron (with a single negative charge) and the proton (with a positive charge). Because protons are concentrated in the nucleus of the atom, the nucleus itself has an overall positive charge which, in a non-ionized atom, is balanced by an equal negative charge contributed by the electrons orbiting the nucleus.

Ions and subatomic particles attract and repel each other following the rules of static electricity; unlike charges attract each other, and like charges repel.

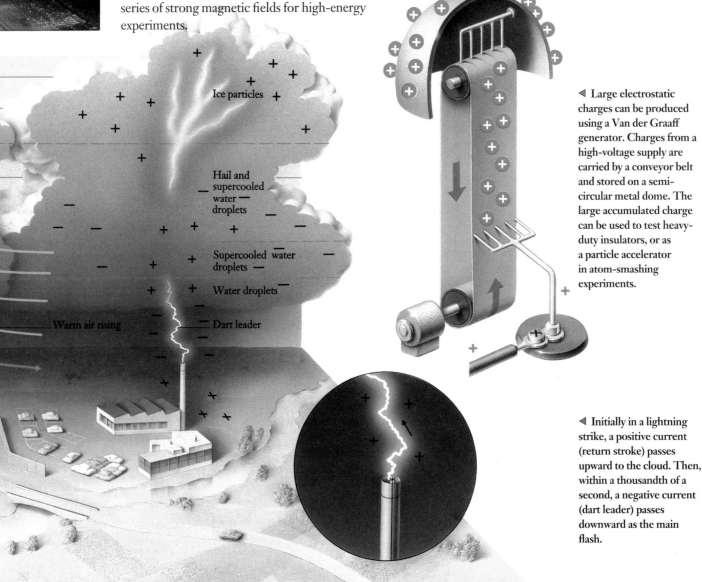

◀ Large electrostatic charges can be produced using a Van der Graaff generator. Charges from a high-voltage supply are carried by a conveyor belt and stored on a semi-circular metal dome. The large accumulated charge can be used to test heavy-duty insulators, or as a particle accelerator in atom-smashing experiments.

◀ Initially in a lightning strike, a positive current (return stroke) passes upward to the cloud. Then, within a thousandth of a second, a negative current (dart leader) passes downward as the main flash.

Ice particles

Hail and supercooled water droplets

Supercooled water droplets

Water droplets

Warm air rising

Dart leader

ELECTRIC CURRENT

AN ELECTRIC current, measured in amperes (amps), consists of a flow of electrons. Some materials are better conductors than others. For a nonmetallic substance such as plastic or glass, it takes a huge potential difference across the material to make the atoms' electrons break away and carry charge. But with metals, even a small potential difference – usually known as a voltage – causes a current to flow. Most metals are good conductors of electricity.

Before 20th-century scientists discovered the key role played by electrons in electricity, they had to assign a direction to current flow, and they established the convention that it flows from a point of positive charge to one of negative charge. In fact, the negatively-charged electrons flow the other way around a circuit, from negative to positive, but the convention about the direction of an electric current has been retained. An atom or molecule that has lost an electron (or electrons) is left with a positive charge and is called an ion. Ions too can be conductors of electric current.

Not all metals conduct electricity equally well, depending on their availability of electrons. The best conductors include aluminum, copper, gold and silver. Because they are less expensive than gold and silver, copper and aluminum are the ones commonly used to make wires and cables for carrying electric current.

The property of a substance that opposes the flow of electricity is known as its resistance. It can be measured by applying a voltage and measuring how much current flows. The German physicist Georg Ohm (1787–1854) established the interrelationship between voltage, current and resistance. Ohm's law states that, for a given material, resistance equals voltage (potential difference) divided by current. The unit of resistance is the ohm, and ohms therefore equal volts divided by amps.

Electricity, because of its ability to move charges, is a form of energy. It can therefore be converted into other forms of energy. For example, when an electric current flows along a piece of wire, the wire is heated. The higher the resistance of the wire, the hotter it gets. If it gets hot enough, it becomes incandescent and gives off light. Electric heaters and electric lamps both have coils of wire that produce heat or light in this way.

KEYWORDS

ELECTRIC CHARGE
ELECTRIC CURRENT
ELECTRIC POTENTIAL
ELECTROMAGNET
OHM'S LAW
POTENTIAL DIFFERENCE
RESISTANCE
SOLENOID
VOLTAGE

▲ Electricity has many applications in industry, homes and leisure activities, making it one of the most useful and versatile forms of energy. Electric lighting was one of the earliest applications and is today probably the most important use of electric current. Many machines, including a fairground carousel, rely on electric motors, and the vast range of electronic devices – from domestic hi-fi to super-computers – depend totally on electricity.

Insulated wire

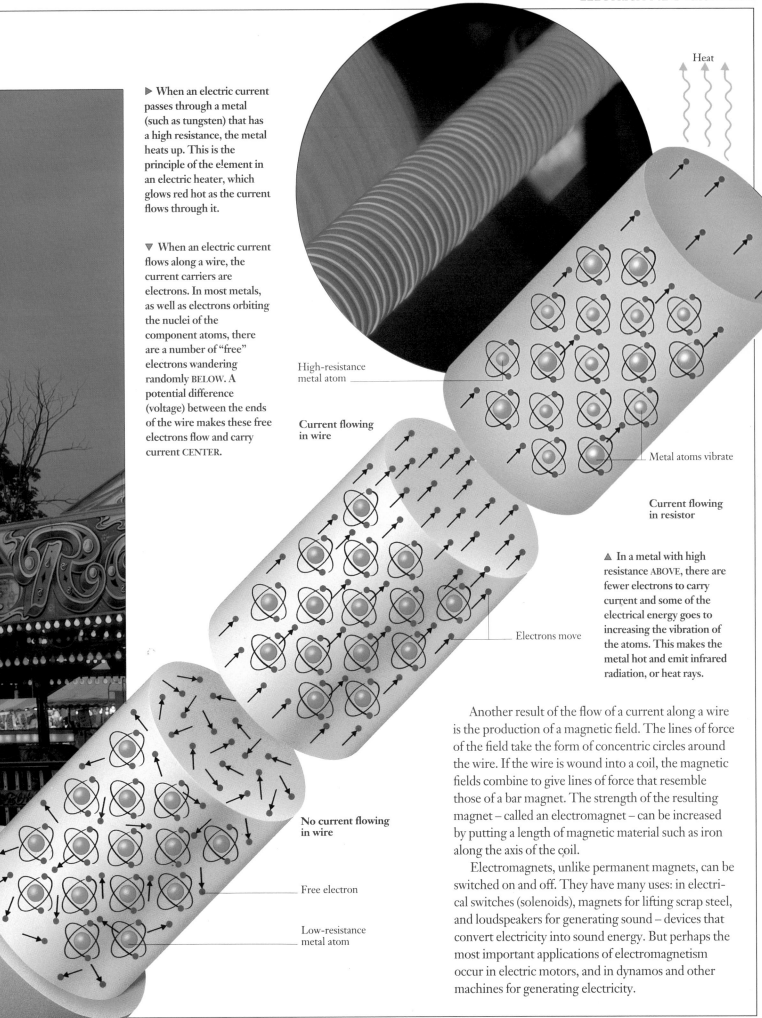

▶ When an electric current passes through a metal (such as tungsten) that has a high resistance, the metal heats up. This is the principle of the element in an electric heater, which glows red hot as the current flows through it.

▼ When an electric current flows along a wire, the current carriers are electrons. In most metals, as well as electrons orbiting the nuclei of the component atoms, there are a number of "free" electrons wandering randomly BELOW. A potential difference (voltage) between the ends of the wire makes these free electrons flow and carry current CENTER.

High-resistance metal atom

Current flowing in wire

Heat

Metal atoms vibrate

Current flowing in resistor

▲ In a metal with high resistance ABOVE, there are fewer electrons to carry current and some of the electrical energy goes to increasing the vibration of the atoms. This makes the metal hot and emit infrared radiation, or heat rays.

Electrons move

No current flowing in wire

Free electron

Low-resistance metal atom

Another result of the flow of a current along a wire is the production of a magnetic field. The lines of force of the field take the form of concentric circles around the wire. If the wire is wound into a coil, the magnetic fields combine to give lines of force that resemble those of a bar magnet. The strength of the resulting magnet – called an electromagnet – can be increased by putting a length of magnetic material such as iron along the axis of the coil.

Electromagnets, unlike permanent magnets, can be switched on and off. They have many uses: in electrical switches (solenoids), magnets for lifting scrap steel, and loudspeakers for generating sound – devices that convert electricity into sound energy. But perhaps the most important applications of electromagnetism occur in electric motors, and in dynamos and other machines for generating electricity.

PRODUCING ELECTRIC CURRENT

ANY DEVICE that makes electrons flow along a wire is a current generator. There are many kinds of generator. Some are chemical, but the most important kind is mechanical and depends on the interaction of electricity and magnetism.

When a length of wire moves in a magnetic field, a current is induced in the wire. If the wire is bent into a coil (or series of coils) and turned in the field between the poles of a magnet, a simple rotating switch can make it produce a continuous flow of electricity. The device for converting mechanical energy into electrical energy – the dynamo – was first conceived more than 150 years ago by the English scientist Michael Faraday (1791–1867).

Thirty years before Faraday's discovery, an Italian physicist, Alessandro Volta (1745–1827), found another way of making electric current. This converts chemical energy into electrical energy, and is commonly called a battery, or cell. In a simple Daniell cell, for example, a piece of copper (called an electrode) dips into a solution of copper sulfate, separated by a porous barrier from the other electrode – a piece of zinc dipping in dilute sulfuric acid.

As the zinc dissolves in the acid to form zinc ions, it releases electrons. These move as an electric current along a wire connecting the two electrodes. Electrons arriving at the copper electrode combine with positive copper ions from the solution to deposit metallic copper on the electrode. The electron producer (zinc) has a negative charge and is called the cathode, while the electron receiver (copper) is the positive anode.

Electrons flow along the external circuit from cathode to anode although, because of the old established convention, current is said to flow the other way (from positive to negative). While electrons are moving along the wire joining the electrodes, ions travel between the electrodes, acting as charge carriers to complete the circuit. This kind of battery is an example of a primary cell.

Dry batteries, as used in flashlights and personal stereos, are also primary cells. When their chemicals are used up, they cease to function. Secondary cells, such as the large battery (or accumulator) that starts a car engine, can be recharged and used again and again. An external voltage applied across the electrodes reverses the current-generating reactions and restores the cell to working order.

The electricity generated by dynamos and batteries is called direct current (DC), because it flows in one direction. That is the purpose of the switching device in a dynamo, but in a machine without one – called an alternator – the current flows first in one direction and then in the other. This kind of electricity, called alternating current (AC), is supplied by the mains to houses and industry. Alternating current is preferred because, for a given voltage, it can be carried along thinner wires and can easily be converted to another voltage using a transformer. The circuits in electronic equipment, such as televisions and hi-fi systems, use direct current. If they are connected to an AC supply, the current is first converted to DC by a rectifier.

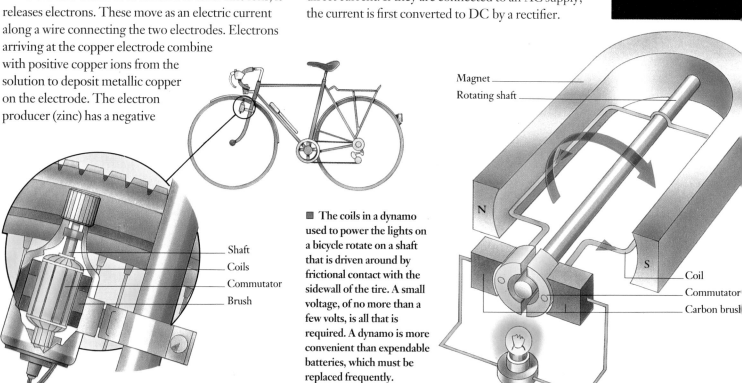

■ The coils in a dynamo used to power the lights on a bicycle rotate on a shaft that is driven around by frictional contact with the sidewall of the tire. A small voltage, of no more than a few volts, is all that is required. A dynamo is more convenient than expendable batteries, which must be replaced frequently.

Shaft
Coils
Commutator
Brush

Magnet
Rotating shaft

N

S

Coil
Commutator
Carbon brush

Cathode
Anode

▲ One of the most convenient portable sources of electricity is the dry battery. This may be used in flashlights, or to power small mechanical devices. The most common design, a type of primary cell, consists of a zinc case (the cathode) holding a paste of ammonium chloride. A central carbon anode is surrounded by powdered manganese dioxide, and the whole cell is capped with a waterproof seal. Dry batteries made from nickel and cadmium can be recharged. They are dry secondary cells.

▲ Using batteries is one way to produce electric current where there is no mains electricity supply and no easy access to sunlight strong enough to power photovoltaic cells. Miners (as here), cavers and divers all use battery-powered lamps. Such lamps are also used by firefighters and others who must work in places where the electricity supply has been cut off.

◀ The principle of the dynamo is best explained by a single coil attached to a shaft rotating in a magnetic field. The commutator keeps current flowing in the same direction.

▶ The battery used in cars and trucks is a secondary cell called an accumulator because it can be recharged. In a car, the generator recharges the battery continually. The usual design of an accumulator contains electrodes ("plates") of lead and lead oxide in a strong sulfuric acid solution. The acid molecules divide into hydrogen and sulfate ions; in a circuit the positively charged hydrogen ions are attracted to the anode, and the sulfate ions to the cathode. The potential of a single lead-acid cell is about 2 volts, so accumulators have six cells to produce the 12-volt output required.

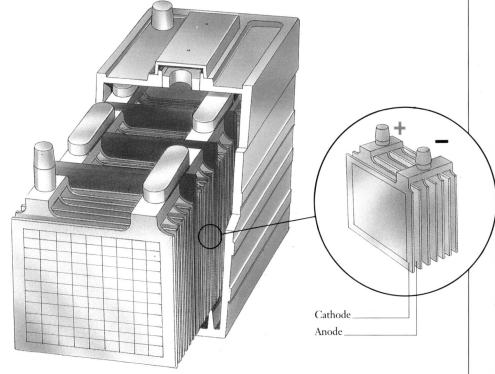

+   −

Cathode
Anode

GENERATING ELECTRICITY

Electricity is an extremely useful form of energy because it can conveniently be distributed to wherever it is required, and once there, it is easy to convert into other forms of energy — that is, to do work. The commercial generation of electricity is a major industry. It has to make use of other forms of energy, which are first converted into mechanical energy to drive electric generators.

Often the prime energy is heat from a fossil fuel such as coal, oil or natural gas, which is burned to generate steam for driving turbines (which turn the generators). Alternatively, heat from a nuclear reactor can be used to make steam for steam turbines. Flowing water, often coming from behind a dam, can be used to turn water turbines to drive generators. On a much smaller scale, the fuel may be burned in gas turbine engines to power the generators.

KEYWORDS

ALTERNATING CURRENT
ALTERNATOR
GAS TURBINE
NUCLEAR REACTOR
STEAM TURBINE
TRANSFORMER

Whatever the prime energy source, the electricity from a power station comes from large alternators. They produce alternating current (AC) at a frequency of 50 or 60 hertz (cycles per second), at very high currents and at voltages of a few hundred volts.

Large currents need thick conductors or they get hot and melt. To avoid using heavy cables for major transmission lines, the supply is transformed to lower currents at much higher voltages (300,000 to 400,000 volts). It is transformed down again to 33,000 or 11,000 volts for local distribution, before finally being converted to the mains voltages of 240 or 110 volts for supply to houses and industry. The output from a power station is measured in watts (equal to the product of the current and voltage). A station with a capacity of several hundred million watts can supply the needs of a small town.

Key stages in electricity distribution rely on transformers to raise and lower voltages. A simple transformer consists of a core of soft iron wound with two overlapping coils of insulated wire. Alternating current flowing in the first, or primary, coil acts like an electromagnet to create a rapidly varying magnetic field in the core. This varying field induces an alternating current in the secondary coil.

If there are more turns of wire in the primary than in the secondary, the voltage is reduced (a step-down transformer); if the secondary turns outnumber the primary turns, the voltage is increased (step-up

▶ Whether the initial source of power is falling water RIGHT, or wind BOTTOM, directly driving turbines, or nuclear FAR RIGHT or fossil fuels BELOW which are burned to create the steam that drives the turbines, a power station converts rotary motion into alternating current (AC), which is converted to very high voltages for distribution through a national grid network. At substations, the current is transformed to the lower voltages needed in homes and industry.

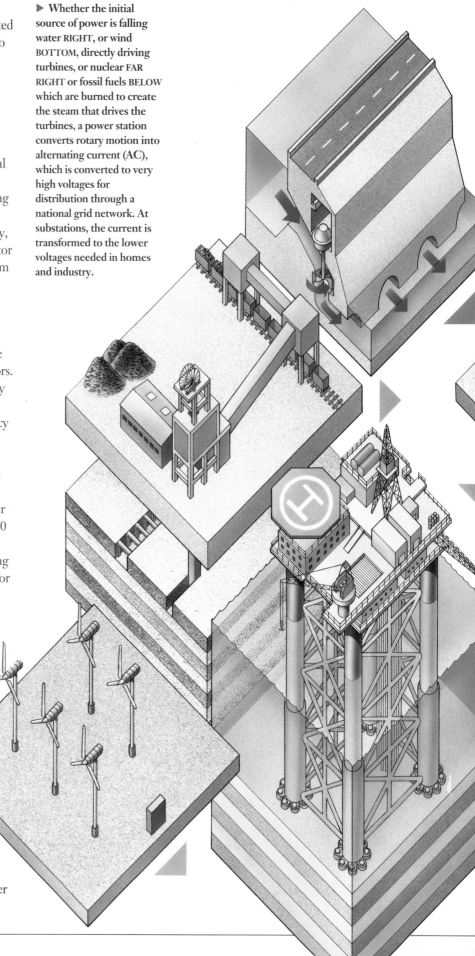

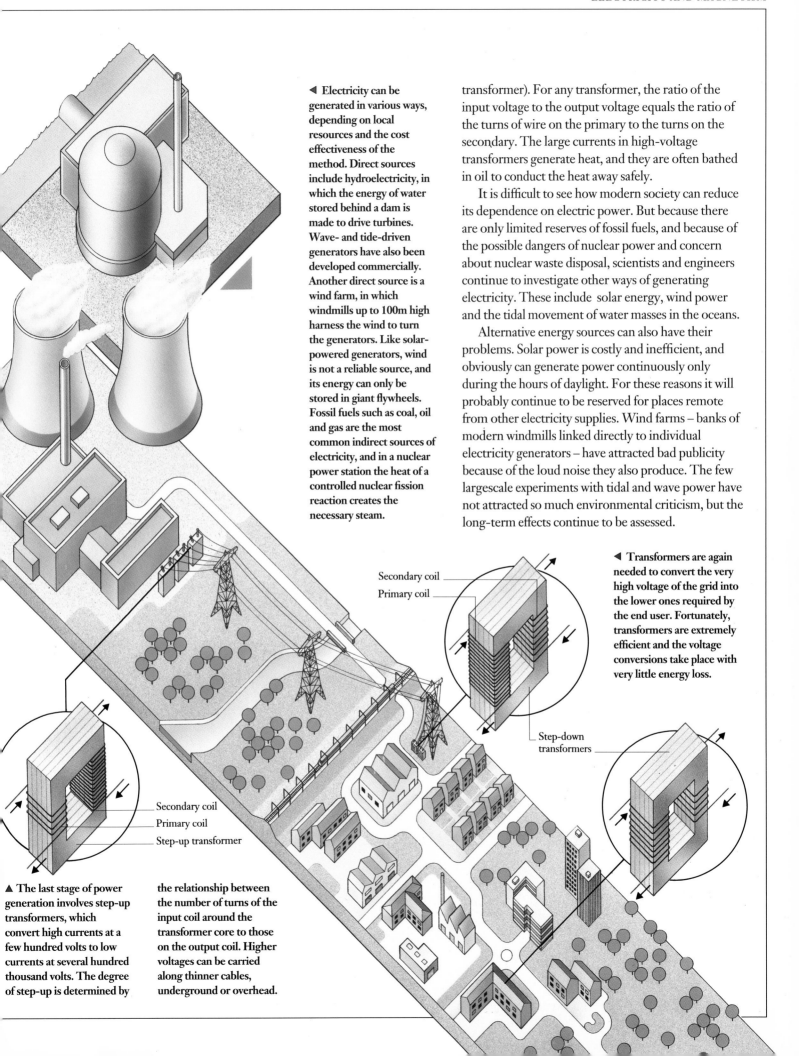

◀ Electricity can be generated in various ways, depending on local resources and the cost effectiveness of the method. Direct sources include hydroelectricity, in which the energy of water stored behind a dam is made to drive turbines. Wave- and tide-driven generators have also been developed commercially. Another direct source is a wind farm, in which windmills up to 100m high harness the wind to turn the generators. Like solar-powered generators, wind is not a reliable source, and its energy can only be stored in giant flywheels. Fossil fuels such as coal, oil and gas are the most common indirect sources of electricity, and in a nuclear power station the heat of a controlled nuclear fission reaction creates the necessary steam.

transformer). For any transformer, the ratio of the input voltage to the output voltage equals the ratio of the turns of wire on the primary to the turns on the secondary. The large currents in high-voltage transformers generate heat, and they are often bathed in oil to conduct the heat away safely.

It is difficult to see how modern society can reduce its dependence on electric power. But because there are only limited reserves of fossil fuels, and because of the possible dangers of nuclear power and concern about nuclear waste disposal, scientists and engineers continue to investigate other ways of generating electricity. These include solar energy, wind power and the tidal movement of water masses in the oceans.

Alternative energy sources can also have their problems. Solar power is costly and inefficient, and obviously can generate power continuously only during the hours of daylight. For these reasons it will probably continue to be reserved for places remote from other electricity supplies. Wind farms – banks of modern windmills linked directly to individual electricity generators – have attracted bad publicity because of the loud noise they also produce. The few largescale experiments with tidal and wave power have not attracted so much environmental criticism, but the long-term effects continue to be assessed.

◀ Transformers are again needed to convert the very high voltage of the grid into the lower ones required by the end user. Fortunately, transformers are extremely efficient and the voltage conversions take place with very little energy loss.

Secondary coil
Primary coil

Step-down transformers

Secondary coil
Primary coil
Step-up transformer

▲ The last stage of power generation involves step-up transformers, which convert high currents at a few hundred volts to low currents at several hundred thousand volts. The degree of step-up is determined by the relationship between the number of turns of the input coil around the transformer core to those on the output coil. Higher voltages can be carried along thinner cables, underground or overhead.

ELECTRIC MOTORS

THE simplest electric machines use electrical energy to do useful work, usually by converting it to mechanical energy. In electric motors, the interaction of magnetic and electric fields produces rotary motion. A small direct current (DC) motor has a U-shaped magnet fitted with pole pieces to produce the magnetic field. In larger motors, the magnetic field is itself produced electromagnetically by current flowing through turns of wire around an iron core.

The current to drive the motor flows around a coil of wire, which is mounted so that it can rotate in the magnetic field. The current enters the coil through a split ring of metal called a commutator. For an extremely simple motor with only a single turn of wire in its coil, the commutator has two segments, each connected to wires leading to or from the coil. Practical motors have many coils forming an armature, and they need a correspondingly larger number of segments in the commutator.

When current flows along a wire in a magnetic field, the wire moves. When current flows around the coil in a motor, it rotates. After part of a turn, the commutator reverses the direction of current flow in the coil, with the result that it keeps rotating. With alternating current (AC), the direction of the current changes continuously and rapidly. For this reason, an AC motor does not need a segmented commutator. But to start the armature turning – and to make sure it turns in the required direction – commercial AC motors have an extra stationary coil. With the windings used to create the motor's magnetic field, the stationary coil creates a secondary magnetic field. This field rotates, pulling the armature around with it.

The most common type of small AC motor – an induction motor – has no commutator at all. The armature coils are replaced by a set of bars of aluminum or copper joined at each end to a metal ring and embedded in a cylinder of iron. This arrangement forms the armature and, because of its shape, it is called a squirrel cage. A series of

stationary coils mounted cylindrically around the armature – called field windings – create a magnetic field which cuts the cage's metal bars and induces a current in them. This current causes the turning effect common to all electric motors, and the armature rotates inside the outer windings, or stator.

Similar field windings can be incorporated into a long flat stator, and the squirrel cage rotor can also be "opened out" to form a flat armature resting on it. When an AC current flows in the stator, the armature moves sideways, producing a linear motor. Such motors are used on a small scale to move sliding doors, and on a large scale can power a fast, silent linear motor train.

A simpler device – a solenoid – also uses electric current to produce sideways movement. It consists of a cylindrical coil of wire, which behaves like a bar magnet when current flows in it. There is a piece of iron (also called an armature) along the axis of the coil and, when the current is switched on, the magnetic field moves the iron sideways. The moving armature can strike the chimes of an electric doorbell, or open and close the contacts of a switch. In this way, a small current can be made to switch a much larger one, and solenoids are common in switch gear for controlling high currents.

▼ The principles of electric motors are best illustrated by simple machines having only one loop of wire in their coil. On the left BELOW is a DC motor; an AC motor is shown on the right. In both machines, electric current is supplied to turn the motor, which is shown driving a pulley. In the DC motor, the current passes to a pair of carbon brushes which route it to a commutator with two segments. The commutator reverses the direction of the current each half turn, to keep the coil spinning within the magnetic field in the same direction. The direction of an alternating current (AC) changes rapidly 50 or 60 times a second, and for this reason the AC motor does not need a commutator.

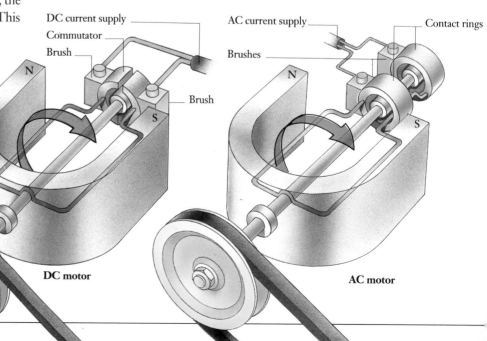

DC current supply
Commutator
Brush
N
S
Brush
DC motor

AC current supply
Brushes
N
S
Contact rings
AC motor

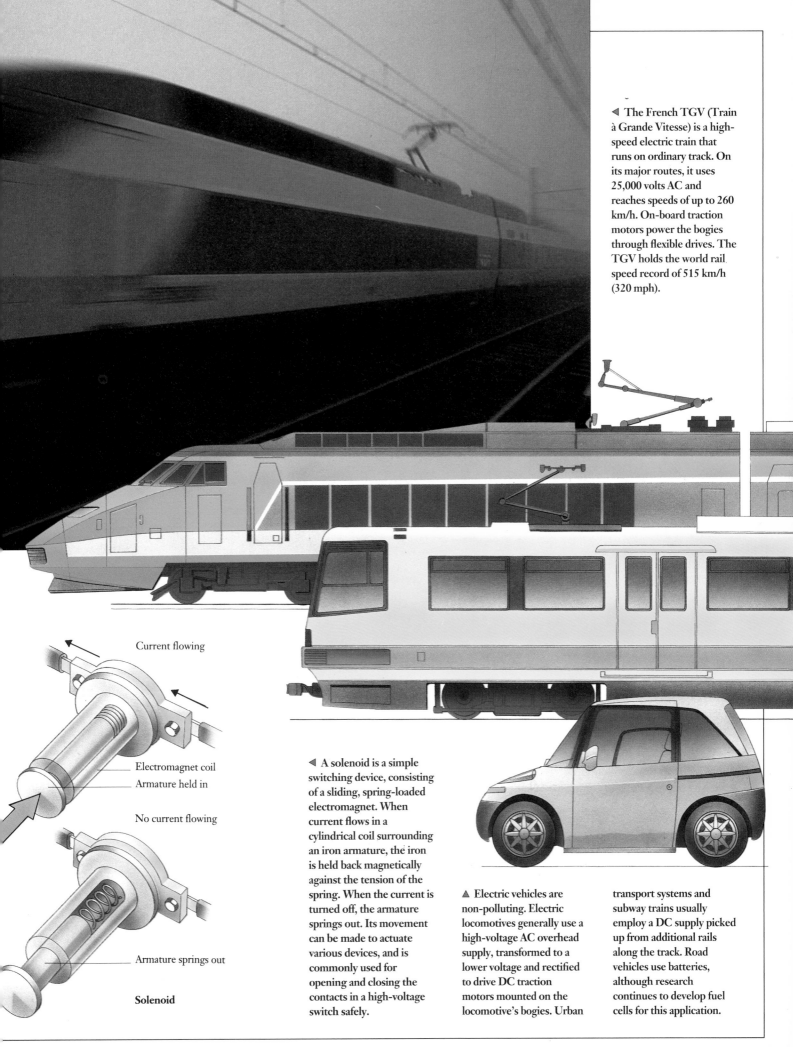

◀ The French TGV (Train à Grande Vitesse) is a high-speed electric train that runs on ordinary track. On its major routes, it uses 25,000 volts AC and reaches speeds of up to 260 km/h. On-board traction motors power the bogies through flexible drives. The TGV holds the world rail speed record of 515 km/h (320 mph).

Current flowing

Electromagnet coil

Armature held in

No current flowing

Armature springs out

Solenoid

◀ A solenoid is a simple switching device, consisting of a sliding, spring-loaded electromagnet. When current flows in a cylindrical coil surrounding an iron armature, the iron is held back magnetically against the tension of the spring. When the current is turned off, the armature springs out. Its movement can be made to actuate various devices, and is commonly used for opening and closing the contacts in a high-voltage switch safely.

▲ Electric vehicles are non-polluting. Electric locomotives generally use a high-voltage AC overhead supply, transformed to a lower voltage and rectified to drive DC traction motors mounted on the locomotive's bogies. Urban transport systems and subway trains usually employ a DC supply picked up from additional rails along the track. Road vehicles use batteries, although research continues to develop fuel cells for this application.

ELECTROLYSIS

I N A BATTERY, chemical energy is converted into electrical energy – a chemical reaction generates electric current. Conversely, electricity can be used to bring about a chemical reaction. This is the principle underlying electrolysis.

A simple example is the electrolysis of water (containing a small quantity of an acid to make it a better conductor). If two pieces of metal, called electrodes, are dipped into acidified water and a battery is connected between them, bubbles of gas form on the electrodes. Water is a chemical compound of hydrogen and oxygen with the well-known formula H_2O. The electricity has the effect of splitting the water molecules in two, so that hydrogen and oxygen gases are evolved at the electrodes.

Many other substances can be decomposed in a similar way by electrolysis, particularly salts in solution or in the molten state. This is because molten salts and solutions of salts are dissociated into ions, which are charged either positively or negatively. The amount of a substance released during electrolysis depends on the quantity of electricity used. This relationship between the two is one of Faraday's laws of electrolysis, named for their discoverer, the English chemist and physicist Michael Faraday (1791–1867).

Electrolysis is one of the most important commercial applications of electricity, after power and light. Two major industrial applications of electrolysis are in the extraction of elements from their compounds and in electroplating. Chlorine gas is manufactured by the electrolysis of a solution of sea water or common salt (sodium chloride, NaCl). Pure copper is produced by the electrolysis of a solution of copper salts, and reactive metals such as aluminum and magnesium are obtained by electrolysis of their molten ores. Often the high temperature needed to melt the ore is provided by an electric arc furnace.

In most electroplating, the object to be plated – which may range from a bolt or steel plate to the tip of a drill or elegant tablewear and jewelry – is made the cathode of an electrolytic cell. The anode consists of the plating metal or an inert metal such as stainless steel. The electrolyte contains a salt of the plating metal. When current flows through the cell, ions of the metal travel to the cathode, become discharged, and are deposited as a coat upon the object to be plated. Electroplating is commonly employed for protective and decorative finishes, using a range of

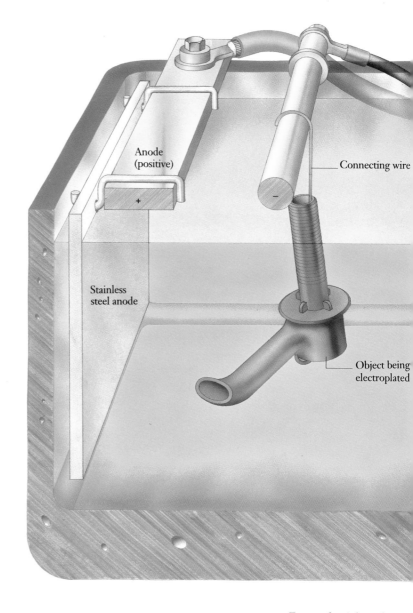

Anode (positive)

Connecting wire

Stainless steel anode

Object being electroplated

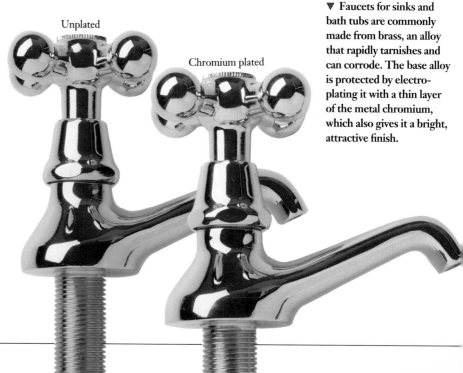

Unplated

Chromium plated

▼ Faucets for sinks and bath tubs are commonly made from brass, an alloy that rapidly tarnishes and can corrode. The base alloy is protected by electroplating it with a thin layer of the metal chromium, which also gives it a bright, attractive finish.

metals, such as cadmium, chromium, copper, nickel, tin, silver and gold.

In the applications described so far, the substance to be extracted or the object to be electroplated forms the cathode (negative electrode) of an electrolytic cell. But anodic reactions can also be used. For example, if an object made of aluminum or one of its alloys is made the anode (positive electrode) in an electrolyte consisting of a strong alkali such as caustic soda (sodium hydroxide), the object is given a thin, adherent coat of oxide. This process is called anodizing. The oxide layer protects the metal against abrasion and corrosion and, because of its chemical nature, can also be dyed or printed to produce various colors or designs on the metal. Steel and various copper alloys such as brass and bronze can also be anodized in order to give them decorative finishes.

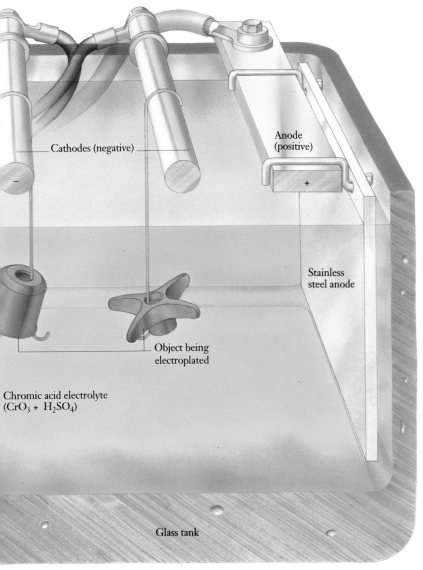

Cathodes (negative)

Anode (positive)

−

+

Stainless steel anode

Object being electroplated

Chromic acid electrolyte ($CrO_3 + H_2SO_4$)

Glass tank

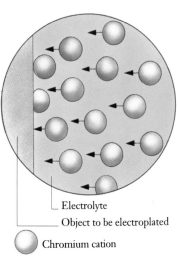

Electrolyte

Object to be electroplated

Chromium cation

◀ Electroplating makes use of electrolysis, in which a direct electric current is carried through a solution (an electrolyte) by the passage of charged ions. Positively charged ions (cations) are attracted to the negatively charged electrode (cathode), whereas negatively charged ions (anions) move toward the positive anode. In one type of chromium plating, the electrolyte consists of chromic acid (a solution of chromium trioxide in strong sulfuric acid) and the anodes are made of stainless steel. The articles to be plated are connected to the negative electrodes and a large direct current is passed through the electrolyte. At the cathodes, chromium cations are discharged and deposited as chromium metal on the articles being plated. Today most metals – even alloys – can be electroplated to impart corrosion resistance and to give a decorative finish. Electroplating was one of the first industrial applications of electricity to be developed; it was used commercially as early as the 1840s by British industrialists for silver plating and gilding.

▶ For a high-quality chromium finish on steel, the object to be plated RIGHT is first coated with copper and nickel. If it is plated directly onto steel FAR RIGHT, pinholes in the chromium admit water, the steel rusts, and the plating flakes off.

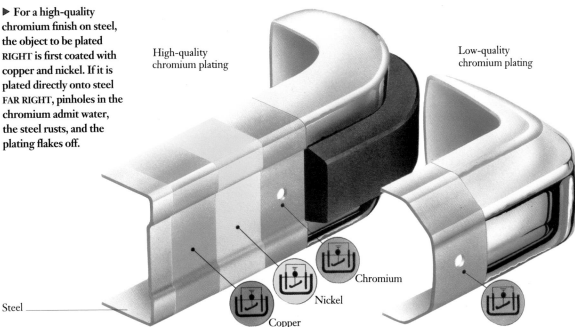

High-quality chromium plating

Low-quality chromium plating

Steel

Copper

Nickel

Chromium

ELECTRICITY AND OTHER ENERGY

Because various forms of energy can be changed into each other, electricity can be generated directly from light or heat. It can even be produced from mechanical energy, without the use of electromagnetism as in a dynamo or alternator. When light or other kinds of electromagnetic radiation such as ultraviolet or X rays shine onto a metal, electrons are emitted from its surface. This is called the photoelectric effect, and the electrons flow to an anode (a piece of metal at a higher voltage than the emitter) and then in an external circuit as an electric current. The photoelectric effect arises when the incoming light has enough energy – that is, it is of a high enough frequency – to knock electrons out of the metal's atoms.

The photoelectric effect is put to good use in photocells and solar panels. A photographer's exposure meter, for example, may contain a selenium photocell and a sensitive current detector (ammeter); the more brighter the light, the greater is the current produced. Most solar panels use photoresistors or photovoltaic cells made of a semiconductor such as silicon. Electricity made this way is expensive, but may be the only practical option for applications such as a vehicle in outer space. Solar panels are employed in sunny climates to generate electricity. They are particularly useful far from an electricity supply.

Thermoelectricity – generating electricity by heat – is produced by joining two wires of different metals to form a loop, and keeping the junctions at different temperatures. Called the Seebeck effect, after the German physicist Thomas Seebeck (1770–1831), it arises because electrons exist at different energy levels in the atoms of the dissimilar metals. At the junctions, electrons flow from one metal to the other. The greater the temperature difference between the junctions, the greater is the current flow. By keeping one junction at a known temperature, the effect can be used to make a thermometer for measuring the temperature at the other junction.

When current passes through a wire in a vacuum, the wire gets hot and gives off streams of electrons. Called the thermionic effect, this is the source of

■ A cheap and efficient method of converting light directly into electricity could solve many of the world's energy problems. Photocells and solar panels, which make use of the photoelectric effect, achieve this conversion, but they are very expensive and not very efficient. Nevertheless, in outer space and in areas remote from a reliable source of electricity, photo-electric devices can provide the only reliable source of electricity – as long as the Sun is shining.

▼ The stylus (or "needle") in some record-player pick-up cartridges consists of a crystal. The V-shaped grooves on an LP stereo audio disk are contoured on each side of the "valley". The undulations are analog versions of the varying sound signals. As the disk turns, the stylus follows the pair of undulations and is thereby vibrated in two directions at right angles to each other. The vibrations cause a small magnet mounted on the crystal to oscillate between pairs of coils. The magnet's movements induce varying electric currents in the coils, one signal for each stereo channel. The signals are then amplified and filtered, and pass to loudspeakers to reproduce the original sounds. Overall, the process has converted vibrations in the stylus into electrical energy, and then back to sound energy in the form of vibrations in the speakers.

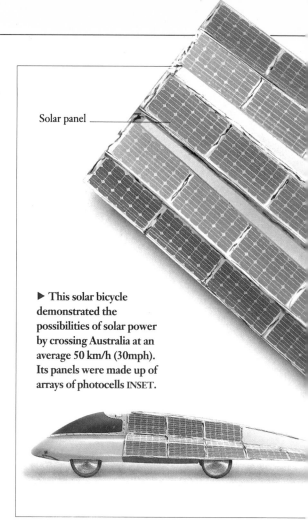

Solar panel

▶ This solar bicycle demonstrated the possibilities of solar power by crossing Australia at an average 50 km/h (30mph). Its panels were made up of arrays of photocells INSET.

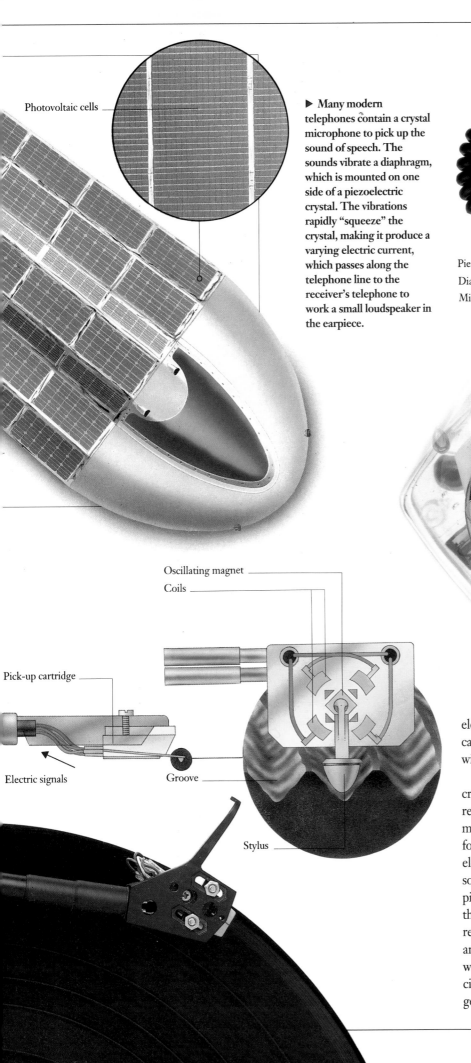

Photovoltaic cells

▶ Many modern telephones contain a crystal microphone to pick up the sound of speech. The sounds vibrate a diaphragm, which is mounted on one side of a piezoelectric crystal. The vibrations rapidly "squeeze" the crystal, making it produce a varying electric current, which passes along the telephone line to the receiver's telephone to work a small loudspeaker in the earpiece.

Piezoelectric crystal
Diaphragm
Microphone cover

Oscillating magnet
Coils

Pick-up cartridge

Electric signals

Groove

Stylus

electrons in a thermionic valve (vacuum tube) or cathode-ray tube, which often has a heated cathode, with a stream of electrons flowing to an anode.

The "needle" on a record player contains a small crystal, usually of sapphire or diamond. When the record turns, a mechanical force causes the crystal to move up and down in the groove of the record. The force "squeezes" the crystal, making it generate a small electric current (which is amplified to generate sounds). This electricity generation – called the piezoelectric effect – occurs because opposite faces of the deformed crystal acquire opposite electric charges, resulting in a flow of electrons. Piezoelectric crystals are also used in some microphones (in which sound waves exert pressure on the crystal) and in "electronic" cigarette lighters (in which a crystal is squeezed to generate current that causes a spark to ignite the gas).

ELECTRONICS AND SEMICONDUCTORS

SOME unusual ways of generating electricity involve the flow of electrons in a vacuum and in solids other than metals. Electronic devices – as they are known – are used to switch and control the flow of currents carrying information, such as sound signals in an amplifier or the digital data signals in a computer.

The first electronic devices were valves (vacuum tubes), in which a stream of electrons flowed from a heated cathode to an anode. This property is used in a two-electrode tube, or diode, for changing AC current into DC current. Adding a third electrode, or grid, makes a triode, which can be used for controlling and amplifying current flow. Heated cathodes are still used in cathode-ray tubes in television sets, radar and computer displays.

But vacuum tubes are bulky and their heaters consume power. With the development of computers requiring complex circuitry, in the years following World War II, the need arose for smaller electronic devices. Coincidentally, just at this time, scientists in the United States invented the transistor, a solid-state device equivalent to the triode. The term solid state means that the electrons travel only in solid substances, not a gas or a vacuum. A transistor consumes no power and can be made extremely small.

A semiconductor is a material that has an electric resistance less than an insulator but greater than a conductor. A metal has many free electrons in its structure which can move from atom to atom in order to conduct current; an insulator has hardly any. A semiconductor, such as the elements germanium and silicon, has some free electrons, which can act as current carriers. Both of these elements have four outer electrons in their atoms. The addition of a very small amount of another element with five outer electrons, such as phosphorous – a process called doping – provides extra conducting electrons, creating an *n*-type semiconductor. Doping with an element with three outer electrons, such as boron, leaves some atoms with a deficiency of electrons (called holes) into which free electrons can flow. The resulting material is called a *p*-type semiconductor. Joining a piece of *n*-type

semiconductor to a piece of *p*-type semiconductor creates a diode which conducts electric current in only one direction. Free electrons from the *n*-type flow across the junction to occupy holes in the *p*-type, but cannot flow from the *p*-type to the *n*-type.

Two junction diodes back to back (an *n-p-n* or *p-n-p* arrangement) form a transistor. A small current fed to the middle piece (the base) controls a larger current between the outer pieces (the emitter and the collector). It acts as a triode and can be used in amplifiers and other circuits. In a field-effect transistor, one type of semiconductor (the gate) is diffused into the sides of a rod of the other type. The main current is fed between the ends of the rod (at the source and the drain). A smaller varying current supplied to the gate controls the main current as the base current controls the emitter in a junction transistor.

■ The earliest electronic devices were vacuum tubes, first with two electrodes (diodes), then with three electrodes (triodes) or more. But tubes are large and their heaters consume power. Their modern equivalents, semiconductor diodes and transistors, are much smaller and consume little or no power. Today's microminiature circuits house hundreds of components on a single silicon chip.

■ Early valve radios LEFT – using vacuum tubes – were bulky and needed heavy transformers to produce the low-voltage current required for their heaters. Transistors RIGHT made electronic devices portable.

◄ Even without the loudspeaker, a valve radio takes up a lot of space. The valves heat up in use, and air circulates to cool the device. Most of the small components and wiring are under the chassis.

▲ A modern personal stereo radio cassette player incorporates a complete FM radio and cassette player in a case small enough to hold in the palm of the hand. A radio receiver can be built much smaller than this, limited only by the size of the loudspeaker. The electronics themselves are incorporated on one or two microchips.

DIODES AND TRANSISTORS

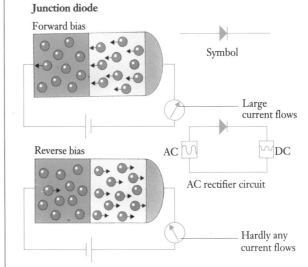

Semiconductor materials

n-type

p-type

Hole

Electron

Junction diode

Forward bias

Symbol

Large current flows

Reverse bias

AC

DC

AC rectifier circuit

Hardly any current flows

Junction transistors

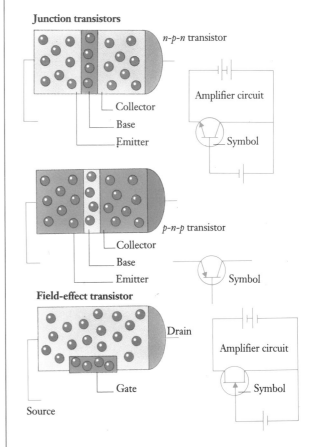

n-p-n transistor

Collector

Base

Emitter

Amplifier circuit

Symbol

p-n-p transistor

Collector

Base

Emitter

Symbol

Field-effect transistor

Drain

Amplifier circuit

Gate

Symbol

Source

MINIATURIZATION

ONE of the great advantages of solid-state electronic devices is their small size. Soon after the development of the transistor in the late 1940s, engineers began to design equally small capacitors, inductors and resistors – the other main components in any electronic circuit. The first consumer product of the new technology was the transistor radio set, which amazed people by its small size.

Thousands of such circuits are needed to build a computer. For the earliest computers the task of mounting the various components and wiring them together was very time-consuming. The wiring problem was solved with the invention of printed circuit boards, which consist of ribbons of copper foil glued to a plastic base. Components are soldered into holes drilled through the copper and the plastic.

Printed circuit board manufacture begins with a sheet of copper-clad plastic laminate, or "board". The copper is coated with a photographic emulsion, called a resist. The circuit design is produced in negative as a series of clear tracks on black film, placed in contact with the now light-sensitive copper, and exposed to strong light. When the exposed board is "developed" (usually by washing in running water), the resist washes away except from the areas – the tracks – that were exposed to light. The board is then placed in a tank of etching solution, which dissolves away all the copper except where it is protected by the resist from the action of the etchant. When the resist is finally abraded away, the printed circuit remains as shiny copper tracks on the board.

Similar photographic techniques were later adapted to make microminiature circuits. In these, the components and their interconnections are formed by building up or etching away consecutive layers of semiconductor material, insulator or metal deposited on a thin slice or "chip" of silicon. Called integrated circuits, they find their most common application in computers. For example, a typical RAM (random access memory) cell has four conducting or semiconducting layers separated by three insulating layers, all on a substrate of silicon or the newer alternative semiconductor gallium arsenide.

The greatest impact of integrated circuits – silicon chips – has been in computers, particularly in the size and capacity of microprocessors, which can be

KEYWORDS

CAPACITOR
INDUCTANCE
INTEGRATED CIRCUIT
PHOTORESIST PROCESS
PRINTED CIRCUIT
RESISTANCE
SEMICONDUCTOR
SILICON CHIP
SOLID-STATE PHYSICS

▷ Modern electronic circuits bring together printed circuit technology with the miniaturization and high component packing density that can be achieved on silicon chips. As a result, manufacturers can produce a mainframe computer the size of a large cupboard, whereas a computer of equivalent power would have occupied a large room only 15 years earlier, and (if it could have been built) would have been the size of a small city using the technology available to the computer-builders of the 1940s.

▷ The design for a printed circuit is first drawn and photographed to produce a negative 1. Copperclad plastic board is coated with light-sensitive emulsion 2, dried, and exposed to light with the negative in contact 3. The exposed board is treated to remove the unaffected emulsion and then immersed in an etchant 4 to dissolve most of the copper. Only the circuit tracks remain 5, from which the emulsion is removed before the chips and circuit components are mounted onto it 6.

△ Rows of screens linked to mainframe computers display data in a tele-communications control room. Such massed electronics has only become possible with the miniaturization of the circuitry.

◁ In manufacturing a silicon chip, several hundred identical chips are produced on a 10 cm diameter, 0.5 mm-thick wafer of silicon UPPER LEFT. Fewer than a quarter of all the chips on the wafer may be of usable quality; the defective ones have to be sorted out by eye. Circuit connections etched out of aluminum foil are then fixed in position before individual chips are split off and mounted on a module LOWER LEFT which is finally encased in its plastic housing for installation.

programmed for any given function. In 1971, the first microprocessors on the market contained 2000 transistors, wired together on printed circuit boards. By the 1990s there were single microprocessor chips that each contained up to a million transistors. Using the technique called very large scale integration (VLSI), such chips confer on a computer the crucial benefit of high speed of operation, making possible the modern generation of microcomputers. One VLSI device is called a transputer (transistor + computer), which is a complete computer on a single chip – there may be more than 25,000 components on a chip only 10 millimeters square. The chip can be linked to other transputers to build very powerful microcomputers.

Silicon chips also have many applications other than within computer circuitry. Today they are found in everything from clocks and watches, pocket calculators, and controllers for domestic appliances to electronic tape measures, personal stereos (tape or CD) and engine management systems for cars. All work through the control of small electric currents – flows of electrons – in solid-state devices. Each contains one or more microprocessors, so that instead of having to design a new chip for each application, engineers install a standard chip that is already programmed for a particular task.

4

SOUND
as Energy

SOUND IS A FORM OF ENERGY that originates when something vibrates – such as a guitar string, the human vocal cords or the reed in the mouthpiece of a saxophone. The vibrations cause waves of alternate high pressure (compression) and low pressure (rarefaction) in the molecules of the air. Sound waves travel out in all directions from their source at a speed of 334 meters per second in dry air. They also travel in other mediums, such as water and solids, and the denser the medium, the faster they travel. If there is no medium to carry them, sound waves cannot move; sound will not travel through a vacuum.

As with all forms of wave motion, sound has wavelength – the distance between the crests (or troughs) of consecutive waves. The number of waves generated each second is its frequency, typically between 20 and 20,000 hertz (cycles per second) in the range audible to humans. Frequency of a pure tone is a measure of its pitch. The amplitude of a sound wave is its height above mean level and is a measure of intensity.

Intensity is not exactly the same as loudness, because the loudness of sound depends also on its frequency. Loudness is a measure of the amount of sound power that passes through a particular area each second. This is measured in decibels.

The vibrating strings of the violins in an orchestra send waves of changing pressure through the air, causing the sensation we call sound. In fact, all sounds are caused by vibrations – such as those of the skin of a drum, the metal of a cymbal or a column of air inside a wind instrument. Sound travels faster through a dense medium than through a less dense one, but in the absence of any medium (as in a vacuum), sound does not travel at all.

PRODUCING SOUND

ROM an engine rattle in a car to the whistling of a draft through a window, sound is produced by vibrations. Musical instruments make use of this fact. In a percussion instrument such as a drum or a cymbal, a plastic skin or thin sheet of metal vibrates and generates sound when struck. In a violin and similar stringed instruments, sound is generated when the strings are kept vibrating by the action of a bow. The body of a violin or guitar acts as a resonator, vibrating at the same frequency as the basic tone generated by the string and so amplifying it. The shape of the body ensures that it resonates at most frequencies within its range.

KEYWORDS

DECIBEL
FREQUENCY
INTENSITY
INTERFERENCE
LOUDNESS
PITCH

The pitch (frequency) of sound from a vibrating string depends on three properties of the string: its thickness, its length and its tension. A string that is thick, long or slack produces a lower tone than one that is thin, short or taut. High notes on a guitar or violin are played on the thinner strings. To get even higher notes, a guitarist (or violinist) presses the string to the fingerboard, making the string shorter. The change in pitch with tension can be heard when a string player tunes the instrument by increasing or decreasing the tightness of the strings.

The sound from a flute, trumpet or other wind instrument comes from a vibrating column of air. The air takes the form of a "standing wave", with alternate nodes (at which the air is stationary) and antinodes (at which it has maximum vibration). Higher notes are produced by blowing harder, creating standing waves with a greater number of nodes and antinodes. Alternatively, the player can uncover holes or press valves, making the vibrating air column shorter.

When sound waves strike an obstruction, they bounce back. In a very large room or hall – such as a cathedral – sounds bounce off the walls and ceiling. This results in reverberation, in which the listener hears the same sound at several slightly different moments, depending on how far it has traveled as it bounces around. Concert halls are specially designed using sound-absorbing surfaces to minimize reverberation, so that every member of the audience hears the sounds at the same time.

If the reflecting surface is large and more than about 30 meters away from the sound source, the reflection is heard as an echo. With a small obstacle, or at the edge of a building, sound waves merely bend around the obstruction in a process called diffraction. This phenomenon explains how sound travels around

■ All sounds, from a rock band to an opera singer's highest note, begin as regular vibrations – as with the strings of a guitar or the vocal cords of a human. Noise is also sound, but unlike music, it is a mixture of sound without any definite pitch.

corners. The direction of sound waves also changes when they travel from one medium to another of different density – an effect known as refraction.

Two sound waves of about the same frequency can combine to produce a new wave. Called interference, this phenomenon gives rise to regular variations in loudness known as beats. When the peaks of the two combining waves coincide, a louder sound is created. When the peaks of one wave coincide with the troughs of the other, the two sounds cancel each other. The number of beats heard per second equals the difference between the frequencies of the combining waves.

Musicians make use of beats when tuning an instrument to a standard reference note (an orchestra tunes to the note A played on an oboe). The instrument is "in tune" – playing exactly the same A as that of the oboe – when no beats can be heard. A tuning fork – a metal implement with two prongs – may also be used to produce a standard tone to which the tuning of an instrument may be matched.

▶ In a wind instrument such as a saxophone, the player vibrates a reed in the mouthpiece which, in turn, vibrates the column of air in the body of the instrument, making a "standing wave" which remains stationary in the body. The standing wave or vibrating air column produces sound waves, which consist of alternate compressions and rarefactions of air radiating out of the bell of the instrument. Other musical instruments produce sounds from vibrating strings (piano, harp, guitar, violin family), skins (drums) or metal (cymbals, bells).

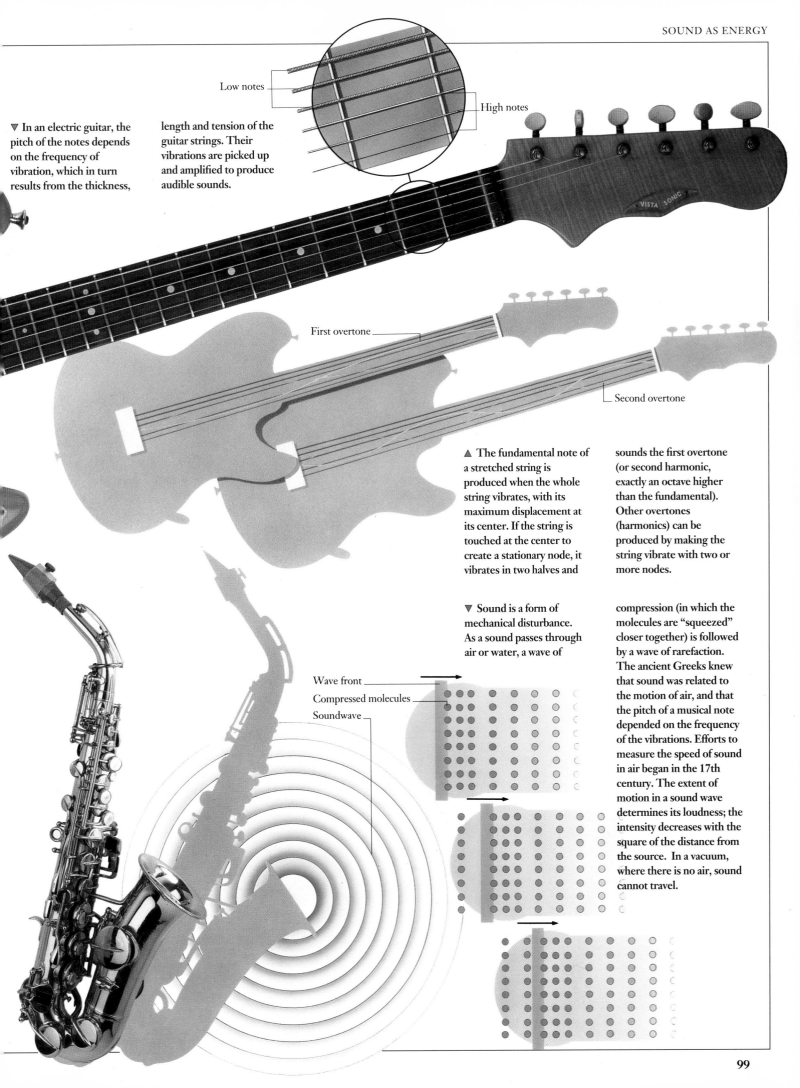

Low notes

High notes

▼ In an electric guitar, the pitch of the notes depends on the frequency of vibration, which in turn results from the thickness, length and tension of the guitar strings. Their vibrations are picked up and amplified to produce audible sounds.

First overtone

Second overtone

▲ The fundamental note of a stretched string is produced when the whole string vibrates, with its maximum displacement at its center. If the string is touched at the center to create a stationary node, it vibrates in two halves and sounds the first overtone (or second harmonic, exactly an octave higher than the fundamental). Other overtones (harmonics) can be produced by making the string vibrate with two or more nodes.

▼ Sound is a form of mechanical disturbance. As a sound passes through air or water, a wave of compression (in which the molecules are "squeezed" closer together) is followed by a wave of rarefaction. The ancient Greeks knew that sound was related to the motion of air, and that the pitch of a musical note depended on the frequency of the vibrations. Efforts to measure the speed of sound in air began in the 17th century. The extent of motion in a sound wave determines its loudness; the intensity decreases with the square of the distance from the source. In a vacuum, where there is no air, sound cannot travel.

Wave front

Compressed molecules

Soundwave

SPEED OF SOUND

IN DRY air at sea level, sound travels at a speed of 334 meters per second. The speed of sound increases with temperature and with altitude, because the air is less dense at higher temperatures and altitudes. But in a very dense medium, such as metal or glass, sound travels up to 15 times as fast as it does in air. A large difference in density prevents sound from passing from one medium to another. Double-paned glass is an effective sound insulator for this reason: hardly any of the sound that passes through the first sheet of glass crosses the layer of air to the second sheet.

Strange effects occur when the source of a sound is supersonic, meaning that it travels faster than the speed of sound itself. When a supersonic jet aircraft flies overhead, the series of compressions in the air (caused by the engine noise) forms a shock wave that follows the aircraft and creates a loud bang called a sonic boom as it passes a listener. It is not an isolated bang but part of a continuous noise which trails in the wake of the aircraft. The wave of compression first forms in front of the aircraft as it approaches the speed of sound, creating shock waves that caused wings of early subsonic aircraft to fail.

This "sound barrier" proved to be an obstacle for aircraft designers until it was first broken by an American rocket-powered aircraft in 1947. The sound barrier consists of a "wall" of high-pressure air caused by the buildup of sound waves, and extra force is needed to enable an aircraft to break through the barrier. The shock wave that accompanies a sonic boom also represents a sudden increase in pressure, which can be enough to damage structures on the ground in its path.

If the source of a sound source moves, the frequency (pitch) of the sound is affected. When the source of sound rapidly approaches a listener, the sound waves are "squeezed" together, increasing their frequency and raising the pitch of the sound. When the source is moving away from a listener, the waves are "stretched" further apart, and the frequency and pitch fall. This phenomenon is called the Doppler effect, for the Austrian physicist Christian Doppler who first explained it in 1842. It can be heard (for example) as the rise and then fall in pitch of the sound of a motorcycle engine as it speeds past. The Doppler effect is also shown by other wave phenomena (as in

KEYWORDS

DOPPLER EFFECT

FREQUENCY

INFRASOUND

MICROWAVES

PITCH

RADAR

SOUND BARRIER

SUBSONIC SPEED

SUPERSONIC SPEED

▶ The speed of sound is much greater in water than in air (about 1540 meters per second in seawater as opposed to 334 meters per second in dry air). Humpback whales use audible and supersonic sound pulses to communicate with each other underwater over long distances – their low-frequency booming can be heard by another whale up to 80 km away.

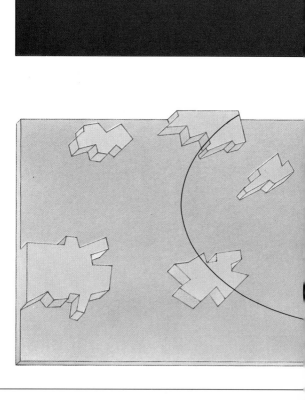

▶ A subsonic aircraft leaves a noise "footprint" on the ground beneath it as it travels along. The movement of the aircraft means that the waves are compressed in front of the aircraft, and extended behind it, causing a change of tone as the aircraft passes. A supersonic aircraft flies faster than the sound waves it generates and thus creates a sonic boom which lags behind it, and is heard only after the aircraft has passed.

Doppler radar) and by light, manifested as the red shift in the spectrum of a fast-receding star.

Ultrasound has frequencies higher than the upper limit of human hearing (about 20,000 hertz for a young adult). It has many applications in technology and nature, as detailed in the article on the following page. Sounds below the lower limit of human hearing (about 20 hertz), called subsonic or infrasound, are used in geological surveying. Engineers detonate explosives underground, and detect the infrasound waves with a series of microphones. The time taken for the waves to reach the microphones provides infor-mation about the types of rocks they traveled through, because rocks bend (refract) the sound waves. In the natural world, creatures as different as spiders and elephants can detect infrasound, and possibly use it as a means of communication.

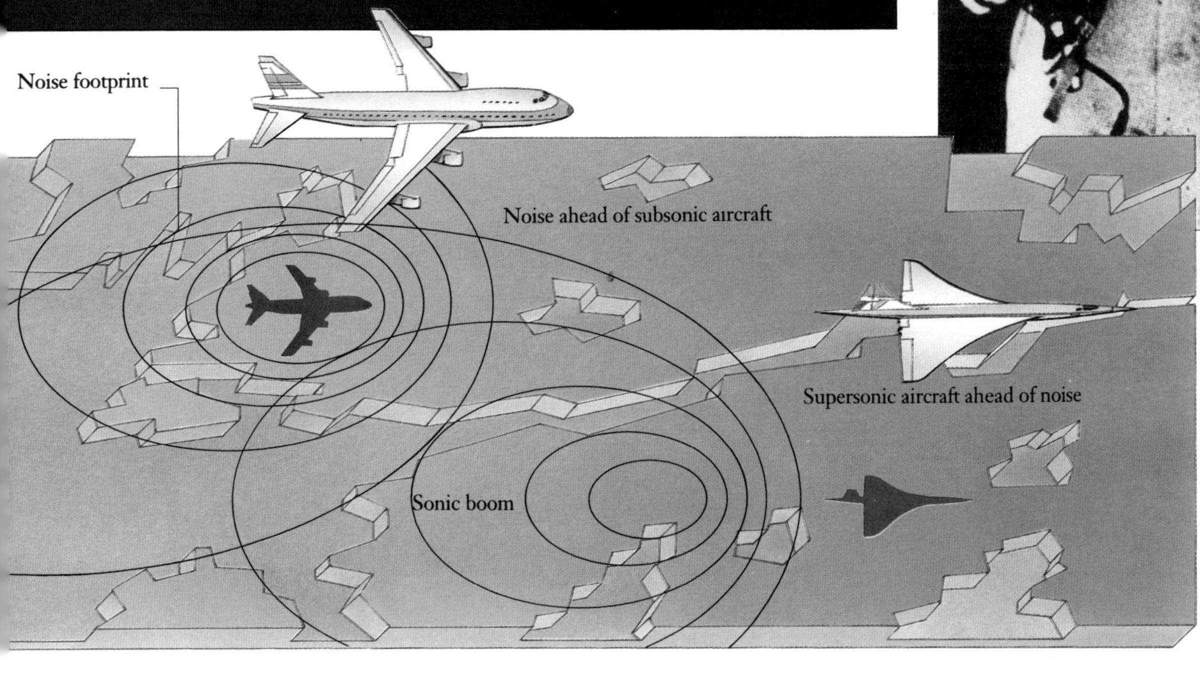

Noise footprint

Noise ahead of subsonic aircraft

Supersonic aircraft ahead of noise

Sonic boom

▲ The sound barrier was broken in 1947 when US pilot Chuck Yeager flew a rocket-powered Bell X-1 aircraft flew at speeds greater than 1150 km/h (Mach 1). A streamlined design minimized the shock effect and buffeting that result when pressure waves build up in front of a subsonic aircraft.

ULTRASOUND

With frequencies greater than about 20,000 hertz, ultrasound is inaudible to humans. But some animals can hear such frequencies and employ ultrasound in ingenious ways. An insect-eating bat, for example, emits ultrasonic squeaks and can hear their echoes. The very high frequency of ultrasound enables the bat to use this form echolocation to detect small objects. As a result, the bat can avoid obstacles in total darkness, or locate its prey. Some cetaceans – whales and dolphins – also use ultrasound for navigation and probably for communication.

These natural systems are imitated by sonar (SOund Navigation And Ranging), which employs ultrasound in the frequency range 100,000 to 10 million hertz under water. Sound travels at about 1500 meters per second in water, more than four times as fast as in air. It is transmitted by a transducer (similar in principle to a loudspeaker) and detected by directional hydrophones (similar to microphones). The direction of an echo gives an indication of a target's bearing, and the time taken for the sound signals to travel out and back can be used to calculate its range, or distance. The range equals a quarter of the speed of sound (in water) multiplied by the time the ultrasound pulses take to travel to the target and back. The bearing and range are displayed on a television-type screen, or processed by a computer.

A simple application of sonar is echo sounding, in which the time taken for sound to be reflected back to a vessel off the bottom of the sea is used to calculate the depth of water below its keel. Commercial fishermen use similar devices called fish finders to locate shoals of fish by their sonar echoes; it is an example of echolocation. A more sophisticated device is a side-scan sonar. It sends a pulsed narrow beam of ultrasonic signals at right angles to the course of the ship carrying it. Any echoes are computer processed line by line to build up a "picture" of the object that is causing the reflections.

Scanning with ultrasound also has various medical applications, in which it is valued as a harmless, non-invasive technique for investigating structures within the body. In Western countries, pregnant women routinely have ultrasonic scans of the uterus to check on the progress of the developing fetus. Other organs that can be scanned include the brain and the heart. An ultrasonic probe is passed at various angles over the surface of the patient's body. The sequences of echoes

▷ Some bats transmit ultrasonic pulses at frequencies in excess of 60,000 hertz. Horseshoe bats emit the pulses through their leaf-shaped noses and collect any echoes with their large ears. They hunt moths in thick foliage, and can tell the difference between a flying insect and a rustling leaf. They also make use of the Doppler effect – a change in echo frequency caused by movement of the "target" – to tell whether the prey is flying towards the bat or away from it.

are processed by a computer to build up a picture of the tissues. The resulting echograms are usually displayed on a television-type screen, which can be photographed to provide a permanent record.

Engineers employ ultrasonic probes to detect flaws in castings and welded joints. The echoes reflected from within the metal structure are converted into signals that can be displayed on a computer screen, to reveal the location of any holes or defects. Ultrasonic echoes are also used by farmers to measure the thickness of fat on animals raised for meat. In industry, ultrasonic cleaning is used to remove all traces of grease from components, particularly before electroplating. The parts to be cleaned are immersed in a container of solvent, and an ultrasonic transducer "vibrates" the liquid to enhance its action.

▷ Many bats emit ultrasound to locate their prey in the dark, using the returning echoes to gauge the range and direction of even small insects.

■ Ultrasound finds a practical application in the sonars called fish finders LEFT, which are used by fishing vessels to locate shoals of fish. In another application, ultrasonic pulses are used to scan body tissues safely and painlessly, such as this scan of a baby in its mother's uterus RIGHT.

SOUND RECORDING

MODERN methods of sound recording involve converting sounds into electrical signals – using a microphone – and storing these signals on tape or compact disc, or using them to form an undulating groove in a plastic audio disk.

There are various kinds of microphones. In one simple type, formerly used in telephones, sound waves vibrate a thin metal diaphragm. The movements of the diaphragm in turn compress granules of carbon, changing their resistance. This varying resistance causes similar variations in the voltage across the granules, and the varying voltage becomes an electrical signal corresponding to the sound waves.

A moving coil microphone has a small wire coil attached to a diaphragm. Movements of the diaphragm move the coil between the poles of a magnet, inducing a varying current in the coil. In a crystal microphone – the most common type – sound pressure waves "squeeze" a piezoelectric crystal, which generates a voltage when stressed in this way.

Whatever the type of microphone, its output is a varying voltage. In a tape recorder, the recording head is basically an electromagnet supplied by an amplified version of the microphone's output voltage. The resulting varying magnetic field at the head magnetizes metal oxide particles – usually iron or chromium oxide – in the coating on the tape. The sound is thus recorded as a pattern of magnetized particles. During playback, the tape runs past a "read" head consisting of a coil of wire in which the magnetized particles induce a varying voltage. In this way, the microphone's output is recreated, and passes via an amplifier to a loudspeaker, to be converted back to sound.

For recording on a compact disk, the microphone's continuously varying (analog) signal is first converted into a digital one. The digital signal controls a gas laser that cuts through a resist on a blank glass disk, which is then etched to form a "groove" more than 5 kilometers long consisting of a series of microscopic pits. This glass master disk is used to make a metal sub-master for stamping out compact disks. On playback, a beam of laser light is modulated by reflection off the pits, and these digital signals are reconverted into a varying voltage.

A vinyl audio disk, or LP record, is manufactured in a similar way by stamping from a submaster. But in this case the original master is formed by making the microphone's varying voltage drive a cutter that carves an undulating groove on a blank disk. When the record is played on a record player, the undulations in the groove exert a varying pressure on a piezoelectric crystal in the needle, or vibrate a magnet within a needle coil. The output from the needle – again a varying voltage – is amplified and fed to a loudspeaker.

The recorded sound is reproduced by the loud-speaker, driven by the voltage output of the tape, compact disk or record player. A conventional speaker has a small coil of wire suspended between the poles of a permanent magnet. The coil is also mounted at the center of a diaphragm. When the varying voltage flows through the coil, it moves rapidly back and forth in the field of the magnet. The back-and-forth movements vibrate the diaphragm, generating sound.

A totally different type of sound recording and playback is used for motion pictures. In this system, the sound to be recorded is made to expose a transparent line, or track, of varying width along the edge of the film. When the film is projected, the intensity of a light shining through the track varies with the track width. This light is detected by a photocell and converted to a varying current.

KEYWORDS

ANALOG-TO-DIGITAL
AUDIO DISK
COMPACT DISC
GRAMOPHONE
LASER
LOUDSPEAKER
MAGNETIC TAPE
MICROPHONE
PIEZOELECTRIC EFFECT

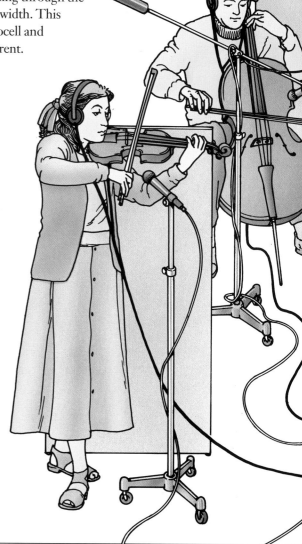

▶ **All forms of sound recording rely on microphones to convert variations in air pressure – sound waves – into electrical signals. Most professional recordings are made using moving-coil microphones. In this type, sound waves vibrate a diaphragm connected to a coil mounted in the field of a permanent magnet. The movement of the coil in the field induces a varying electric current in the coil, which constitutes the audio signal output of the microphone. Other kinds of microphone use carbon granules, piezoelectric crystals or metal ribbons.**

▶ **In recording several voices or instruments, each has its own microphone whose output passes to the control console. Screens may be used to keep the various sounds separate, and the control room of a studio is usually separated by a sound-insulated window. For an orchestra, each section of instruments has its own microphone (with a separate one for any soloist); these separate recordings must be "mixed" by a sound engineer. For a band playing electronic instruments, the audio output from each instrument passes directly to the console. There the sound engineer operates a bank of faders to balance the various outputs, sometimes combining two or more of the signals, before they pass to a multitrack tape recorder.**

▶ The signals from the various microphones in the recording studio are passed via an amplifier to a console, where they can be balanced, then stored on a multitrack tape, with one track for each signal. Later the various tracks are mixed to produce only two – one for each stereo channel – which are stored on a master tape. This tape can then be used to make multiple copies on tape cassette, or to produce a master disk for pressing mass-produced audio disks. If the sounds have been recorded digitally, the master may take the form of a compact laser disk. Vinyl disks and magnetic tape have both conventionally carried analog signals, although digital audio tape was introduced in the early 1990s. Compact discs utilize the digital process, which helps to eliminate errors in signal reading and produce better sound.

▼ At the mixing stage, the sound engineer can work on each track, boosting its volume or tonal range, or removing noise. Multi-channel systems give the listener the feeling of being immersed in the sound.

On/off switch

Moving coil
Magnet
Diaphragm

Microphones
Multi-channel recording

Signals mixed and split into stereo channels
Right channel
Left channel

▼ The chief recording media are audio cassettes, compact discs and vinyl disks, although vinyl is being phased out.

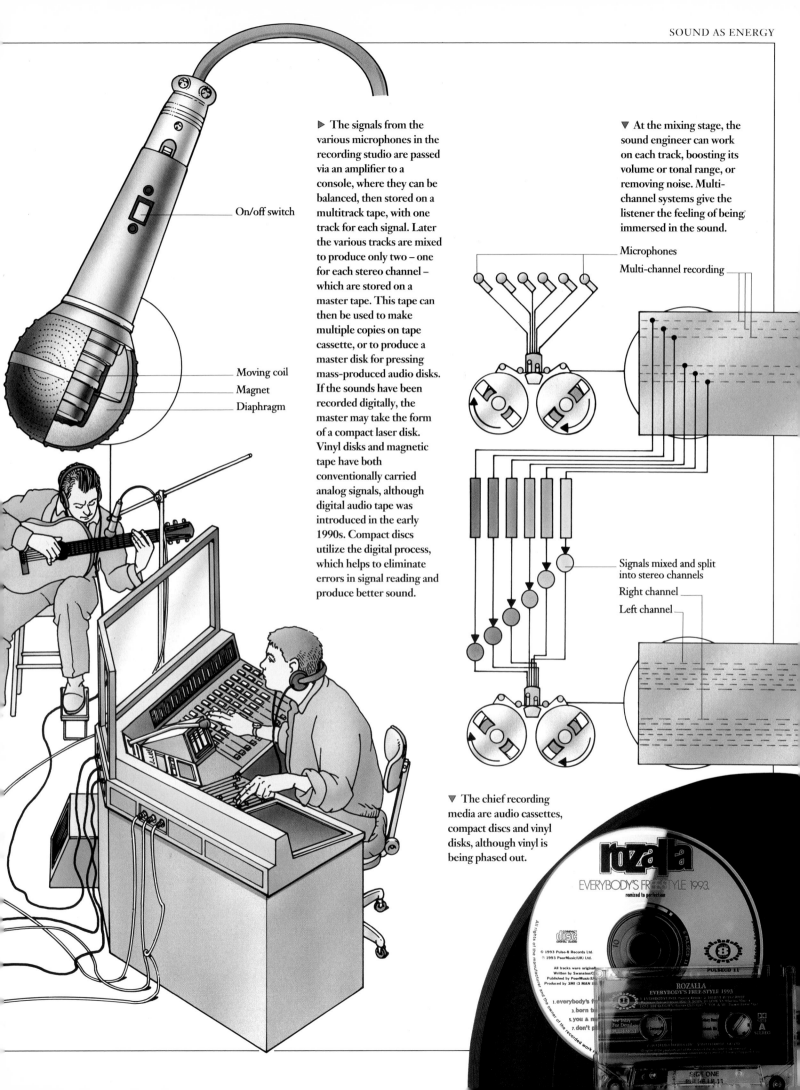

5

LIGHT
& the Spectrum

L IGHT IS A TYPE OF ENERGY that powers all life on Earth. Plants use light from the Sun to build their tissues, and animals – including humans – eat plants or animals that have eaten plants. Every living organism is therefore dependent on light. Light also enables us to see color in the world around us.

Like radio waves and X rays, light is a form of electromagnetic energy. All are produced by the activity of electrons in atoms. Light originates within an atom when some of its electrons are first energized and then lose energy. When electrons "jump" from one energy level to a lower one, the energy difference is given off as radiation – either forms that can be seen (visible light), or beyond the visible spectrum, like infrared, ultraviolet, radio waves and X rays.

Also, like other forms of electromagnetic radiation, light can be described in terms of its frequency and wavelength. Frequency is the number of waves that are generated each second. Wavelength is the distance between successive peaks (or troughs) of a wave.

To the human eye, different light wavelengths appear as different colors. The shorter wavelengths are violet or blue; the longer ones appear as red light. The whole range of visible wavelengths can be seen in a spectrum, which is produced naturally in a rainbow or in the laboratory by passing light through a glass prism.

White light consists of a mixture of colors that are sorted into their various wavelengths when the light is passed through a prism to form a spectrum. The spectrum is produced because the glass of the prism refracts (bends) the light rays. The longer wavelengths, at the red end of the spectrum, are bent least, while violet and blue light have shorter wavelengths, and are bent most.

PRODUCING LIGHT

ALL devices for producing light, from a candle or electric lamp to a fluorescent tube or laser, depend on processes that take place within atoms. All of these processes involve electrons.

In a neutral atom, the electrons occupy different orbits representing different levels of energy: the orbit closest to the nucleus has low energy, while the outer orbits have higher energy. Extra energy supplied to the atom – by heating it, for instance – is absorbed by the electrons and causes them to "jump" to a higher energy level. However, they are unstable in this excited state, and quickly jump back to their original levels. As they do, the extra energy they have absorbed is emitted as light. The wavelength (and therefore the color) of the emitted light varies depending on what element is being used.

Light can be produced in various ways. The chief difference is in the way the extra energy is supplied to the atom. In the flame of a candle or oil lamp, carbon from the hydrocarbons in the wax or oil is heated until it glows. Heat is also the energy source for light from a gas lamp, which has a mantle – a sheath placed around the flame. Atoms of thorium metal in the mantle give off an intense white light.

In an ordinary electric lightbulb, the heat is produced when electric current flows through the thin tungsten wire of the filament; the tungsten atoms give off light. In an electric arc lamp, the intense light comes from a white-hot spark as it jumps between two carbon electrodes.

Another way of converting electron energy into light energy – without involving heat – occurs in a discharge tube, such as the type used for neon advertising signs. The tubes contain traces of neon gas at low pressure. Electric current flowing into an electrode (cathode) at one end of the tube produces a stream of electrons. As these flow along the tube to the electrode (anode) at the other end, they collide with neon atoms, exciting some of their electrons to a higher energy level. When the excited electrons return to their original level, the familiar red neon light is emitted. The use of the gas xenon instead of neon gives a tube that produces the bright white light of a photographer's electronic flash.

■ Most lighthouses (such as the one shown RIGHT from New England) use high-powered electric lamps to produce strong beams that can be seen over long distances. One such type of electric lamp – used for flashing lights – is a form of discharge tube containing the gas xenon. Smaller xenon lamps are used on emergency vehicles and civil aircraft.

◀ An oil lamp has a wick dipped into a container of kerosene. Capillary attraction causes the kerosene to rise steadily up the wick (hidden by the metal casing). When the kerosene is set alight, the carbon atoms in it absorb heat energy and give off light in the flame. Adjusting the height of the wick regulates the size and intensity of the flame.

A fluorescent tube is a slightly different source of unheated (or "cold") electric light. Like a discharge tube, the fluorescent tube also involves an electric current and two electrodes, but in this case the gas used in the tube is mercury vapor at low pressure, which produces invisible ultraviolet light. The inside of the fluorescent tube is coated with a substance called a phosphor. As the coating is struck by ultraviolet light, some of the phosphor's atoms are excited. When they return to their normal stable state, they give off visible light.

Different types of phosphors produce light of different colors. Such phosphors are also used on the inside of a television or computer screen, where they are excited to produce light by streams of electrons in the cathode ray tube.

Phosphors are fluorescent, which means that they stop emitting light after the supply of stimulating radiation (whether ultraviolet light or electron streams) is stopped. A similar phenomenon is known as phosphorescence. In this case, however, the light continues to be emitted for a short time after the stimulating radiation stops. This is how phosphorescent substances such as luminous paints – which absorb sunlight – glow in the dark.

Tube off

Phosphor

Tube on

Heated cathode
Mercury atom
Electron
Ultraviolet light
Visible light

▶ In a fluorescent lamp NEAR RIGHT, a heated cathode emits electrons which collide with mercury atoms. Ultraviolet light from the mercury excites atoms in a phosphor which lines the tube, making them give off visible light. In an ordinary electric lamp FAR RIGHT, light is emitted by the atoms in the heated tungsten filament, held in the inert gas argon.

REFLECTION AND MIRRORS

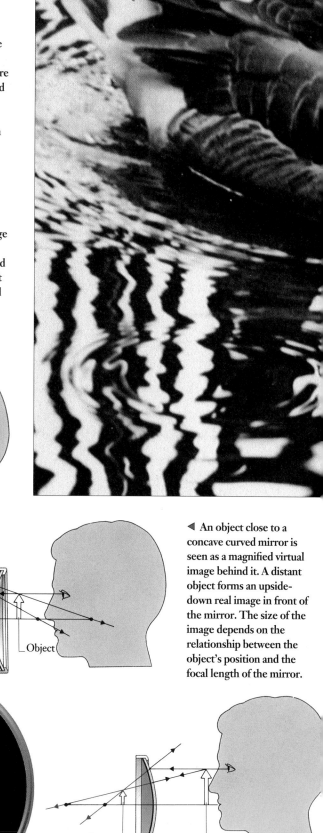

▷ Glass, polished metals and mirrors all reflect light. But the best natural reflector is water, which shimmers in low-angle sunlight or reflects in its ripples the distorted shape of anything floating on it, such as this duck. There are various kinds of mirror and they have a host of uses. Some of the largest and finest mirrors are found in optical telescopes, which employ curved mirrors to focus the faint light from distant stars and galaxies.

N AIR, glass or water, light travels in straight lines. In the branch of physics called optics, which is concerned with the behavior of light through mirrors, lenses and optical instruments, light is often thought of as rays represented by straight lines. Straight rays from a light source such as the Sun or an electric lamp cause an object to cast well-defined shadows.

When light rays strike an object, some of them are reflected; it is the reflected rays that enter our eyes and allow us to see the object. Some materials reflect light better than others. A perfectly black object reflects hardly any light; a highly polished piece of metal reflects nearly all the light rays that strike it. The best reflectors are mirrors, which are usually made by applying a thin coating of silver to a sheet of glass. If a ray of light strikes a plane (flat) mirror at right angles, it is reflected back along the same path, called the normal to the mirror. Light striking a mirror at an angle to the normal is reflected at the same angle on

▽ A plane (flat) mirror produces a same-size image of any object in front of it. It is a virtual image, formed behind the mirror. But left and right are interchanged in the reflection.

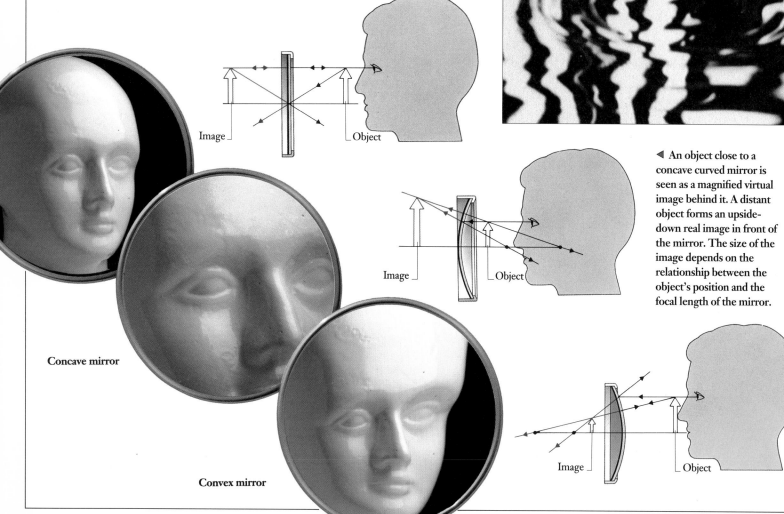

Image   Object

Concave mirror

Convex mirror

Image   Object

◁ An object close to a concave curved mirror is seen as a magnified virtual image behind it. A distant object forms an upside-down real image in front of the mirror. The size of the image depends on the relationship between the object's position and the focal length of the mirror.

Image   Object

the other side of the normal. A law of reflection of light states that the angle of incidence (between the incoming ray and the normal) equals the angle of reflection (between the reflected ray and the normal).

When rays reflected from a mirror reach our eyes, we look back along them as if they had not been reflected and see an image of an object that appears to be behind the mirror. This is called a virtual image, and it is the same distance behind the mirror as the object is in front of it. Plane mirrors are used in combination in periscopes to bend light through two right angles, enabling the viewer to see over obstructions.

Curved mirrors behave differently. There are two types, convex (domed outward toward the viewer) and concave (domed inward, like a saucer). A point on the axis of the mirror at its radius is called the center of curvature; half way between the center of curvature and the mirror is the point called the focal point. Light rays reflected in a convex mirror form a virtual image behind the mirror; the image is smaller than the actual object. Convex mirrors are used as driving mirrors in cars and trucks.

The image formed by a concave mirror depends on the position of the object in relation to the center of curvature and focus. If the object is between the focus and the mirror, a large virtual image is formed behind the mirror. as in the curved mirror used when shaving or putting on makeup. If the object is at the focus, the virtual image is formed an infinite distance behind the mirror, and cannot be seen. When the object is outside the focus of a concave mirror, it produces a real image (so called because the image can be formed and seen on a screen, which is impossible with a virtual image). For an object located outside the center of curvature, the image is upside-down and smaller than life size. But if the object is between the focus and center of curvature, it produces a magnified, upside-down real image.

Light rays originating at the focus of a concave mirror are reflected parallel to the mirror's axis. If a light source, such as an electric lamp, is placed at the focus, the mirror can be used to reflect a parallel beam of light, as in a flashlight or car headlamp. But the major application of curved mirrors is in compact long-focus camera "lenses" and astronomical tele-scopes. A telescope at Mount Palomar in California uses a concave mirror measuring 508 cm (200 inches) across to collect light from distant sources in order to produce images of planets, stars and galaxies. The mirror is more easily supported than a large lens.

▼ A periscope consists essentially of a pair of plane mirrors mounted at 45-degree angles at the ends of a rectangular tube. The image is upright and (unlike an image in a single mirror) not reversed left to right. High-quality periscopes, such as those used on submarines, have silvered 45-degree prisms instead of mirrors, with a system of lenses within the tube that produces magnified images and broadens the field of view. The periscope can also be rotated to give a 360-degree field of vision.

▷ Spectators at a golf tournament use cardboard periscopes to see over the heads of the people in front. By fitting the periscope with curved mirrors, a wider field of view could be obtained. NASA is experi-menting with periscopes on supersonic aircraft to give pilots an improved view below the fuselage.

◁ The image of an object formed by a convex mirror – such as that seen in a car's rear-view mirror – is always upright, reduced in size and virtual (formed behind the mirror).

REFRACTION AND LENSES

WHEN a ray of light passes from one transparent medium to another of different optical density – such as from air to glass – it does not continue to travel in a straight line. On entering the denser medium, the ray is bent away from the normal (the line at right angles to the surface) in the phenomenon called refraction. The amount of refraction depends on the optical densities of the mediums.

The exact behavior of the light ray obeys a law

formulated by the Dutch mathematician and physicist Willebrod van Roijen (1591–1626), who adopted the Latin name Snellius. Snell's law, as it is now known, states that for a light ray of a particular wavelength (color), the sine of the angle of incidence (between the incident ray and the normal) divided by the sine of the angle of refraction is a constant. This constant is the refractive index for the mediums concerned. For example, the refractive index of water is 1.5, and that of crown glass (used in camera lenses) is about 1.3.

Light travels more slowly in a dense medium. Another definition of refractive index equates it to the speed of light in a vacuum divided by the speed of light in the medium concerned. The refractive index of air is virtually the same as that of a vacuum, which is assumed to be 1.

Refractive indices are important in the design and behavior of lenses. There are two basic types: a convex lens is thicker in the middle (like a magnifying glass) and a concave lens is thicker at the edges (like the lenses in glasses for nearsightedness). Light passing

▶ There are two types of lenses: concave (diverging) and convex (converging). Light rays passing through a concave lens diverge, producing a smaller image, as in the artist's reducing glass RIGHT. Through a convex lens, rays converge to a focus, and can produce an enlarged image, as with the magnifying glass FAR RIGHT. Lenses work using refraction, which makes the paintbrush in the glass of water ABOVE appear distorted.

through the exact center – along the axis – of either lens passes right through in a straight line. But light entering a convex lens off its axis is refracted (bent) toward the axis, and refracted again on leaving the lens; thus rays parallel to the axis are brought together at a focus behind the lens. A concave lens refracts light rays away from the axis. Rays parallel to the axis diverge after passing through the lens, and can be regarded as coming from a focus which is on the same side of the lens as the incoming light.

Because of these differences in basic behavior, convex lenses are also known as converging, or positive, lenses, and concave lenses are also called diverging, or negative, lenses. Convex lenses can form real or virtual images, depending on the position of the object in relation to the focus of the lens. Concave lenses always produce virtual images.

The amount a ray of light is refracted by a lens depends on the color, or wavelength, of the light. For example, long-wavelength red light is refracted less than shorter-wavelength blue light. As a result, when white light (which is a mixture of all colors) passes through a simple convex

▶ Droplets of water on the tendril of a plant act as spherical lenses and produce perfect (though inverted) images of the flower behind them.
As with all lenses, the basic cause is refraction – the bending of light rays when they pass from one optical medium to another. The earliest lenses, used for concentrating light from candles and oil lamps, were simply spherical glass flasks filled with water.

lens, its red component is focused slightly farther from the lens than the blue component. An image formed by the lens has colored fringes around its edges, in the phenomenon known as chromatic aberration. This defect is overcome in high-quality lenses by making them of several components, using two types of glass to cancel out the aberration.

Many optical devices make use of lenses. Perhaps the most familiar, apart from the eye, is a camera, in which a convex lens – or a combination of lenses that behave overall like a positive lens – focuses a reduced, upside-down image onto the film. Simple telescopes (sometimes called terrestrial telescopes), binoculars and opera glasses use pairs of lenses to produce magnified images. Greater magnification still is produced by optical microscopes, using a combination of positive lenses.

Unlike terrestrial telescopes, astronomical telescopes using lenses produce upside-down images, which is why most early astronomer's drawings and photographs of the Moon and planets show their north poles at the bottom. The size – and hence the magnification –-of such telescopes is limited by the weight of the lenses and the difficulty of making them accurately, and today's large optical telescopes use mirrors instead of lenses. Mirrors are lighter in weight than lenses, and, if necessary, can be installed in several sections to make installation easier. They can also be adjusted independently.

▼ In a compound microscope, light reflected off a sub-stage mirror illuminates the specimen. The objective lens produces an enlarged image of the specimen, which is further magnified by the eyepiece. The overall magnification is the magnifying power of the objective multiplied by the power of the eyepiece.

▼ The lens in the eye is a convex lens which brings images of objects to a focus on the retina at the back of the eye 1. The image is upside-down, but the brain turns it right way up. A person whose eyeball is slightly too short from front to back is farsighted (hypertropic) because the eye lens tries to focus the light rays behind the retina.

The condition can be corrected by wearing eyeglasses with convex lenses 2. If the eyeball is too long, the result is nearsightedness (myopia), because light is focused in front of the retina. It can be corrected with eyeglasses having concave lenses 3.

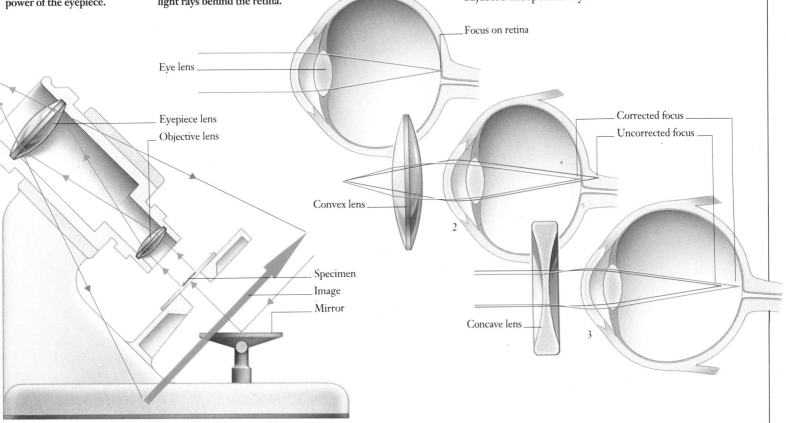

Focus on retina

Eye lens

Eyepiece lens
Objective lens

Convex lens

2

Specimen
Image
Mirror

Concave lens

Corrected focus
Uncorrected focus

3

DISPERSION AND DIFFRACTION

LIGHT rays are bent – refracted – when they pass from one medium to another, such as from air into glass. The extent to which they bend depends on their wavelength, which is related to color: blue light is bent more than red light. For this reason, the image formed by a simple convex lens, for example, is surrounded by colored fringes because the different colors that make up white light are focused at slightly different places. This defect in a lens is called chromatic aberration.

When white light passes through a glass prism, its different wavelengths are bent to different extents both on entering the prism and on leaving it. As a result, the component wavelengths are spread out to form a spectrum ranging from violet and indigo at one end, through blue, green and yellow in the middle, to orange and red at the other end. This phenomenon of spectrum formation is called dispersion. Its most familiar natural example occurs in a rainbow, which forms when sunlight is dispersed and reflected by airborne droplets of rainwater.

Light rays are also bent when they pass through a very narrow slit. But in this effect, called diffraction, red light is bent more than blue light. A useful laboratory tool called a diffraction grating consists of a glass plate ruled with very fine lines spaced 5,000 to 10,000 per centimeter. When a beam of white light is passed through such a grating, it is split into a spectrum. If physicists, astronomers or chemists wish to analyze the spectrum of a particular light source, they use diffraction gratings rather than prisms to create the spectrum.

Other interesting phenomena are possible when light passes through

■ A rainbow RIGHT is caused by the dispersion of light inside raindrops – in much the same way as a glass prism splits white light into a spectrum. But the spectral colors seen, for example, in a peacock's feathers are caused by a different phenomenon: diffraction BELOW. In a primary rainbow FAR RIGHT, raindrops reflect light once before it reaches our eyes at an angle of about 41°. On bright days, there may also be a secondary rainbow outside the primary one. This results from double reflections inside the drops: the light enters our eyes at an angle of about 52°. The order of spectral colors in this bow is reversed.

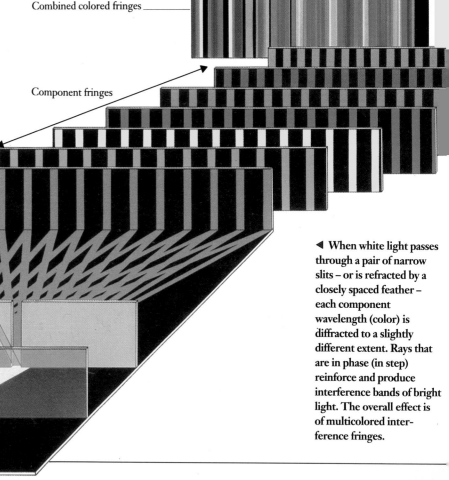

Combined colored fringes

Component fringes

White light

◀ When white light passes through a pair of narrow slits – or is refracted by a closely spaced feather – each component wavelength (color) is diffracted to a slightly different extent. Rays that are in phase (in step) reinforce and produce interference bands of bright light. The overall effect is of multicolored interference fringes.

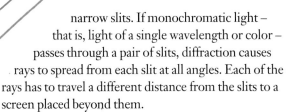

Primary rainbow

Secondary rainbow

52°

41°

Raindrop

Spectral colors

Spectral colors reversed

▼ The colors of a rainbow are produced when a ray of white light from the Sun is refracted (bent) as it enters a raindrop. The ray is then reflected from the back of the drop, before being refracted for a second time as it leaves the drop. Different wavelengths of light (which appear to the observer as different colors) are refracted to a different extent, and so the double refraction has the effect of splitting the white light into a multicolored spectrum that is directed toward the ground.

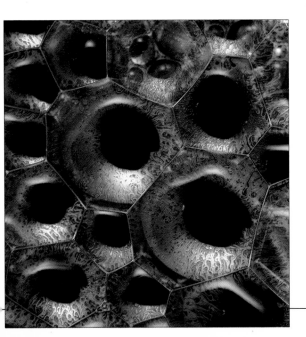

◄ The colors seen in soap bubbles are an interference phenomenon. The interference is caused when rays that are reflected from the front of the soap film interfere with rays that are reflected off the back of the film. A similar effect occurs when daylight reflects off the surface of a thin film of oil floating on water, or when light reflects off the surface of a compact disc.

narrow slits. If monochromatic light – that is, light of a single wavelength or color – passes through a pair of slits, diffraction causes rays to spread from each slit at all angles. Each of the rays has to travel a different distance from the slits to a screen placed beyond them.

If the lengths of the paths traveled by two rays differ by a whole number of wavelengths, they arrive at the screen "in step". They therefore reinforce each other and produce a bright line on the screen. Rays that are out of step cancel each other out in the phenomenon called interference, resulting in a dark band on the screen.

The pattern of light and dark bands formed in this way are called interference fringes. They can be produced in various other ways when light waves are made to travel in and out of step. For instance, a thin film of air trapped between two pieces of glass causes interference when light rays reflected from surfaces at the upper and lower edges of the film differ in path lengths. Concentric fringes produced in this way are called Newton's rings.

Interference also occurs with white light. But in this case the various wavelengths or colors are affected individually, resulting in fringes consisting of all the colors of the rainbow. Light reflected from the upper and lower surfaces of a thin film of oil on water gives rise to colored fringes in this way. The color comes from an optical effect, not from the oil itself.

A similar effect can be seen in soap bubbles, and in the light reflected from the microscopic pits on the surface of a compact disc. It can also been seen in the light reflected from the scales or feathers on the wings of some butterflies and birds.

LIGHT FROM LASERS

A LASER is a device that uses a standard light source to stimulate atoms to produce coherent light (with all the light waves in step); the term laser comes from the initial letters of Light Amplification by Stimulated Emission of Radiation. A simple laser may be based on a cylindrical crystal of ruby, which is silvered at one end to form a mirror. The other end of the crystal is semisilvered or has a central hole, so that it reflects some light and lets some light through.

A flash tube, similar to that in a photographer's flash gun, is coiled around the crystal. When it flashes, its light "excites" some of the atoms in the ruby, causing electrons in the atoms to jump to a much higher energy level. When the flash tube is off, the electrons revert to a lower energy level, but one that is still higher than the original level. Further absorption of light energy by these atoms causes them to emit laser light as the electrons finally return to their original level.

This light is reflected back and forth within the crystal, continuing to stimulate more and more ruby atoms to emit light, while some of it emerges as a pulse of laser light through the semisilvered mirror, or through the hole in one mirror. The ruby laser can produce only short bursts of laser light, but lasers using carbon dioxide or other gases instead of a ruby crystal produce continuous laser light, and the gas atoms can be excited by high-frequency radio waves instead of flashes of light.

Lasers have been used in many applications since they were first produced in the 1960s. In medicine, a laser beam can be used like a fine scalpel to remove skin blemishes and small growths, cauterize broken blood vessels and tack back a detached retina in the eye. Laser beams can be introduced into body cavities along fiber optic tubes. Fiber optics and lasers are also used in telecommunications. Infrared laser beams passing along such tubes are modulated to carry data, telephone signals and television programs – all at once, if required. They use low-powered semiconductor diode lasers, which can also be made small enough to fit inside a portable compact disk player.

A laser beam travels in a straight line, which makes it useful in leveling instruments used in the construction industry. The builders of the tunnel under the English Channel (which was bored simultaneously from each end) used a laser beam to ensure correct alignment of the two halves of the

▼ With the initial flash of intense light from the flash tube, some of the electrons in the laser's ruby atoms are excited to a high energy level 1. These electrons then revert to a lower (but still higher than normal) level 2. At the next flash, they briefly absorb more light and then emit it as coherent laser light when they return to their normal level 3.

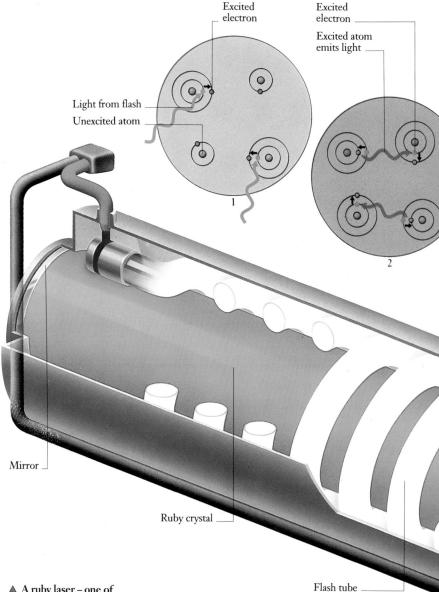

▲ A ruby laser – one of the earliest types to be developed – produces laser light in short pulses, corresponding to the rate of flashing of the flash tube coiled round it. As the light is emitted by the excited atoms in the ruby crystal, it bounces back and forth between mirrors at the ends of the crystal. But one mirror has a hole at its center (or is only semisilvered), and the pulses emerge through this mirror. The inducing flashes are also reflected internally from the mirrors, so that all the excited atoms emit their radiation in phase (with their waves exactly in step) resulting in a pulsed beam of coherent light. The ruby used in lasers is a synthetic form of the mineral corundum (aluminum oxide). The first laser, built by Theodore Maiman in 1960, produced a flash of monochromatic light 10 million times brighter than the Sun.

► A laser beam slicing through the night air above a city provides convincing proof that light travels in straight lines. Such beams have even been sent all the way to the Moon, to be reflected back to Earth by mirrors left there by Apollo astronauts. These have been used to measure precisely the distance to the Moon.

Laser light

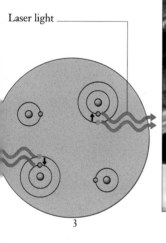

3

Mirror with central hole
Laser beam
Monochromatic light

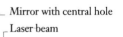

Polychromatic light

Monochromatic and coherent laser light

► Lasers provide surgeons with ultrasharp scalpels. They are used for delicate operations, such as tacking back a detached retina in the eye. Used with fiber optic endoscopes, lasers can also be introduced within the human body.

▲ Lasers produce monochromatic coherent light – that is, all the light is of the same wavelength (and therefore of the same color) and all the waves are exactly in step with each other. This effectively concentrates the energy of the laser light, which can be used for accurately cutting piles of cloth or heavy metals or even for slicing diamonds.

tunnel. Alongside the Bosporus Bridge in Turkey and across the San Andreas Fault in California – both active earthquake areas – there are permanent laser beams aimed at a detector to give advance warning of the slightest earth movements.

Lasers can also be used to produce images called holograms, for storing 3-D graphics and detecting credit card forgeries. To produce a hologram, a laser beam is split into two beams and reflected by a system of mirrors. One beam (the reference beam) shines directly onto a photographic plate. The second beam illuminates an object before reaching the plate, where it combines with the reference beam to produce a complex interference pattern. It is the photographically stored interference pattern that is the hologram of the object. When viewed in laser light, it reveals a 3-D image of the original object.

INVISIBLE RADIATIONS

LIGHT is one type of electromagnetic radiation. Just beyond the visible spectrum, at shorter wavelengths than those of visible light, lies ultraviolet radiation. Ultraviolet light is invisible to human eyes, but can be seen by some insects. At longer wavelengths, on the other side of the visible spectrum, lies infrared radiation. It can be detected by a few animals such as pit vipers. Infrared radiation is given off by anything hotter than its surroundings.

The Sun gives both kinds of radiation as well as visible light. Most of the ultraviolet rays are blocked by the ozone layer in the upper atmosphere. The infrared radiation consists of heat rays that can be felt on Earth, 150,000,000 kilometers away.

The electromagnetic spectrum extends beyond the ultraviolet and the infrared. Shorter wavelengths, in the range 1 to 10^{-6} nanometers, consist of X rays and gamma rays. X rays are produced by changes in metal atoms that have been made unstable by being bombarded by a stream of electrons. In an X-ray tube, a cathode consisting of a wire filament is heated red-hot by an extremely high-voltage electric current (up to 2 million volts). The anode consists of a lump of copper, which often has water pipes incorporated in it to keep it cool. Attached to the copper is a slice of the heavy metal tungsten, termed the target.

Electrons boil off the cathode and stream toward the target, where they excite tungsten atoms, in which electron "jumps" result in the emission of X rays. The rays, emitted at right angles to the electron beam, pass through a window in the side of the X-ray tube. The energy of X rays depends on the voltage applied to the tube. Their chief use is in medicine. They are also used in analytical science.

X rays do not occur naturally on Earth, although they are emitted by certain stars and other celestial bodies. So are gamma rays, which are even more energetic than X rays. They do occur on Earth as an accompaniment to the decay of various radioactive elements, such as isotopes of radium and uranium. Unlike X rays, which arise from excitation of electrons in an atom, they are produced by changes that take place in an atom's nucleus. They are used to take "X-ray" photographs of metallic objects, and to sterilize food and medical equipment.

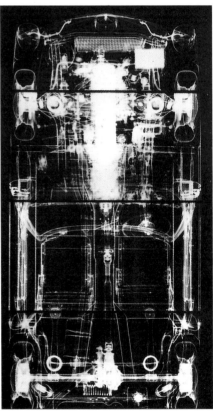

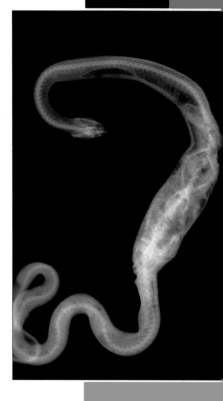

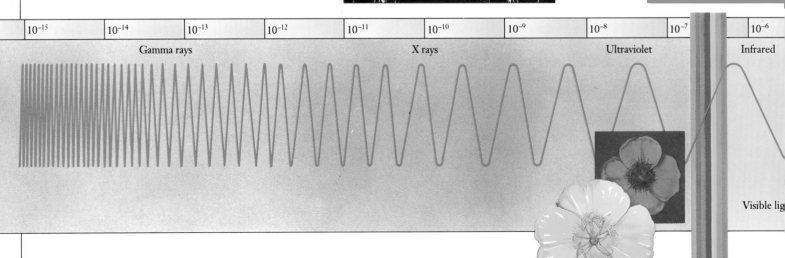

| 10^{-15} | 10^{-14} | 10^{-13} | 10^{-12} | 10^{-11} | 10^{-10} | 10^{-9} | 10^{-8} | 10^{-7} | 10^{-6} |

Gamma rays   X rays   Ultraviolet   Infrared

Visible lig

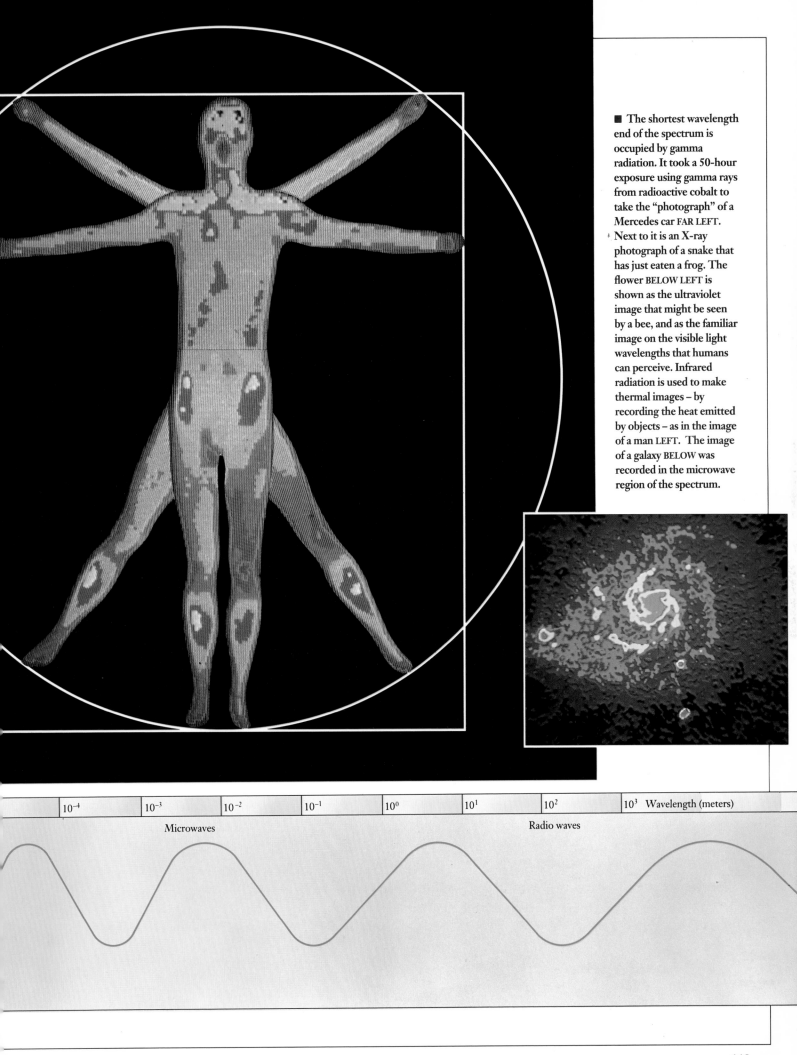

■ The shortest wavelength end of the spectrum is occupied by gamma radiation. It took a 50-hour exposure using gamma rays from radioactive cobalt to take the "photograph" of a Mercedes car FAR LEFT. Next to it is an X-ray photograph of a snake that has just eaten a frog. The flower BELOW LEFT is shown as the ultraviolet image that might be seen by a bee, and as the familiar image on the visible light wavelengths that humans can perceive. Infrared radiation is used to make thermal images – by recording the heat emitted by objects – as in the image of a man LEFT. The image of a galaxy BELOW was recorded in the microwave region of the spectrum.

10^{-4}	10^{-3}	10^{-2}	10^{-1}	10^{0}	10^{1}	10^{2}	10^{3} Wavelength (meters)

Microwaves

Radio waves

RADIO WAVES

THE long-wavelength end of the electromagnetic spectrum is occupied by radio waves. The shortest of these – with wavelengths of 0.1 to 30 cm, just more than those of infrared radiation – are called microwaves. They are used for satellite communications, in radar, for cooking food, and also for direct local radio communications. For longer distances on Earth, microwave signals must be relayed between tall towers that are located up to 50 kilometers (30 miles) apart, the line-of-sight distance.

Microwaves are generated in special electron tubes (valves), in which a high-frequency electric field varies the speeds of streams of electrons. This makes them resonate in a metal cavity, producing micro-waves. A typical microwave transmitting valve, called a klystron, is made of metal and works at very high voltages. The waves are transmitted and received by dish-shaped antennae, which focus a beam of microwaves as a curved mirror focuses a beam of light.

Exploding galaxies called quasars and distant clouds of interstellar gas emit microwaves, which are detected by large radio telescopes. Like other forms of electromagnetic radiation, the waves travel through space at the speed of light. The received signals are extremely weak, but can be amplified by a maser – Microwave Amplification by Stimulated Emission of Radiation. Like the klystron, it too uses a resonant cavity to produce coherent microwave radiation.

Electromagnetic radiation of wavelengths greater than 30 cm are usually known simply as radio waves. They are also produced by oscillating electrons in wires or transmitting valves, and are used mostly for communications. The actual transmitter consists of a metal wire or rod, which sends the radio waves into the air.

Special techniques make the transmitted wave carry data or signals corresponding to speech, music or pictures. The transmitter emits a continuous radio wave, at a particular wavelength, called a carrier wave. Like any other wave, it has a characteristic frequency (number of waves per second) and amplitude (wave height). The signal to be broadcast is made to vary, or modulate, the carrier wave. At the receiver, the broadcast signal is picked up by an antenna and then demodulated – the carrier wave is removed. The remaining audio-frequency signal is amplified and made to work a loudspeaker.

In frequency modulation (FM), the broadcast signal varies the frequency of the carrier wave. In amplitude modulation (AM), the amplitude is varied. FM transmissions use short wavelengths and so, like microwaves, their range is limited to line of sight, and the quality of reception is normally good. AM transmissions may use extremely long wavelengths – up to several hundred meters. They can bounce off layers of ionized gas in the atmosphere and so travel long distances; if strong enough, they can travel around the world. Quality of reception is generally poorer than that of FM because stray signals from electrical machinery or electric storms can interfere with the broadcast signals, producing static.

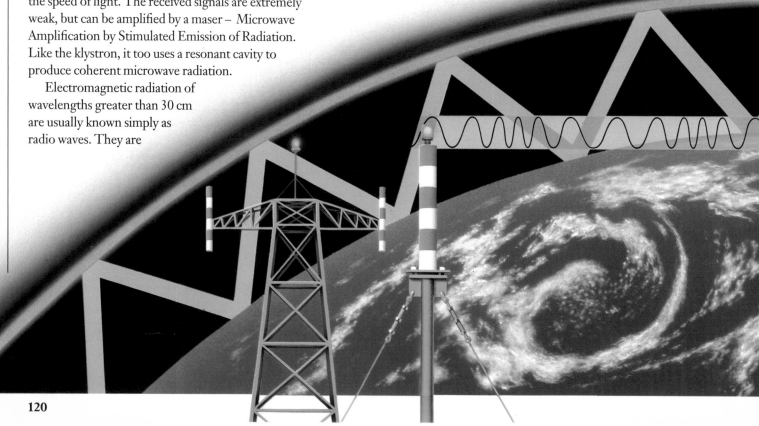

▼ The range of radio waves depends on wavelength. Very high frequency (VHF) microwaves, such as those used for FM broadcasting, travel have an effective range up to 50 km (if the transmitter is on a a tall tower). Microwaves can also be beamed up to a communications satellite and retransmitted to receivers on the ground. Medium wavelength radio waves can bounce off the ionosphere – a layer of ionized gas in the atmosphere – and achieve long ranges by zig-zagging around the globe.

▲ Using computers to analyze microwave signals picked up by radio telescopes, astronomers can construct radio maps of celestial objects such as galaxies. The incoming radio signals are of a particular wavelength – typically about 10 centimetrers – and the computer displays them in various colors, corresponding to the intensity of the signals.

▲ Many objects in outer space, including pulsars and quasars, emit radio waves, usually in the microwave region. They can be received by large dish antennae, known (because of their astronomical function) as radio telescopes. These can be located on Earth because the microwaves – unlike ultraviolet and infrared – penetrate the Earth's atmosphere. Infrared telescopes must be put in space because the rays do not come through.

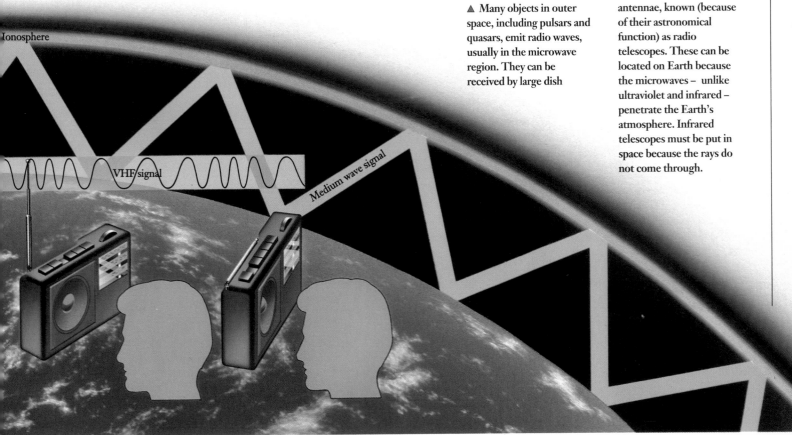

Ionosphere

VHF signal

Medium wave signal

RADAR AT WORK

LONG-RANGE radio communications depend on the reflection of radio waves off the ionosphere – layers of ionized gases in the Earth's upper atmosphere. In the late 1930s, scientists in Britain and Germany independently discovered that large solid objects, such as ships and aircraft, also reflect radio waves in the microwave region. The echoes reflected from such objects, soon to be known collectively as targets, gave information about the direction and distance of the target. The system was developed initially for the military and became known as RAdio Detection And Ranging, or radar.

In a typical radar system, an antenna transmits microwave signals of wavelength 1–10 cm. The signals travel at the speed of light (300,000 kilometers or 186,000 miles per second) to the target, which reflects part of them back to a receiving antenna – which is the same as the transmitting antenna. The echoes are displayed, usually on a television-type screen called a plan position indicator (PPI).

Some transmitting antennas consist of a metal mesh dish or array that can be aimed toward the target. In others, the antenna rotates to make the radar signal search a wide area. In most modern radar devices, this rotation is achieved electronically, and the antenna itself does not move. Because there is only a single antenna, each type of radar has a switching device to make it alternate rapidly between transmitting and receiving.

Most radar systems concentrate the transmitted signals into parallel beams of pulses, rather than continuous waves, and the range of a target is found by measuring the time taken for a pulse to be reflected back. The frequency of the returning signal can also provide information. If the target is moving, there is a change in the frequency of the echo caused by the Doppler effect. If the target is approaching, for example, the reflected microwaves become "squeezed" closer together and their frequency increases. From this increase, the speed of the approaching target can be calculated. For a receding target, the frequency of the reflected wave decreases.

Doppler radars of this type are used by the police to measure the speeds of road vehicles. The Doppler effect also allows continuous-wave radar systems to ignore stationary objects and display only targets that are moving. This suppresses echoes of buildings and hills, preventing clutter from building up on the radar

KEYWORDS

ECHO
DOPPLER EFFECT
ION
IONOSPHERE
MICROWAVES
RADAR

display and highlighting only those targets that are significant.

Astronomers use Doppler radars mounted on satellites to determine the direction and speed of rotation of planets. This is done by pointing the radar at the equatorial edge of a planet to measure its speed of approach or departure. Meteorologists also use satellite radar to plot weather systems – particularly the density of rain clouds – for compiling weather maps and forecasts. Other satellite-borne radar systems accurately measure the heights of mountains on land and under the oceans to construct radar maps of the Earth's surface. Such satellites usually carry a synthetic aperture radar (SAR), which achieves a sideways scan and can produce computer-generated images of mountain ranges and other surface phenomena. They are also equipped with a radar altimeter, which transmits pulses directly downward. The American Seasat satellite used such an altimeter to make detailed maps of the features on the world's ocean floors. Similar radars have even been used to penetrate the dense clouds in the atmosphere of Venus and produce maps of the planet's surface.

▲ In radar systems, a series of microwave pulses is sent towards a target. Any reflections (echoes) are detected on their return. The target's range is measured by the time the pulses take to return.

◀ The common type of radar display is called a plan position indicator (PPI). It displays the outlines of surrounding buildings and geographical features, with any targets as small blips of light that gradually move.

Outgoing pulse

Echo

RADAR MAPPING

Put into orbit by the United States, NASA's Landsat series of earth resources satellites use radars to produce images of the terrain below. Orbiting at 800 km, a Landsat's scanning system covers strips of land measuring 185 km across as it passes overhead. The accumulated data is transmitted to a ground station and processed to produce images like the one below of St Louis, Missouri, during serious flooding in 1993. Similar satellites have been used to make radar maps of other planets, such as Venus.

Landsat

Radar beam

Ground station

◀ The airways above major airports bustle with traffic 24 hours a day. To keep track of the aircraft and prevent collisions, air traffic controllers plot the positions of all aircraft on radar screens, and assign each aircraft an identification number (usually its flight number).

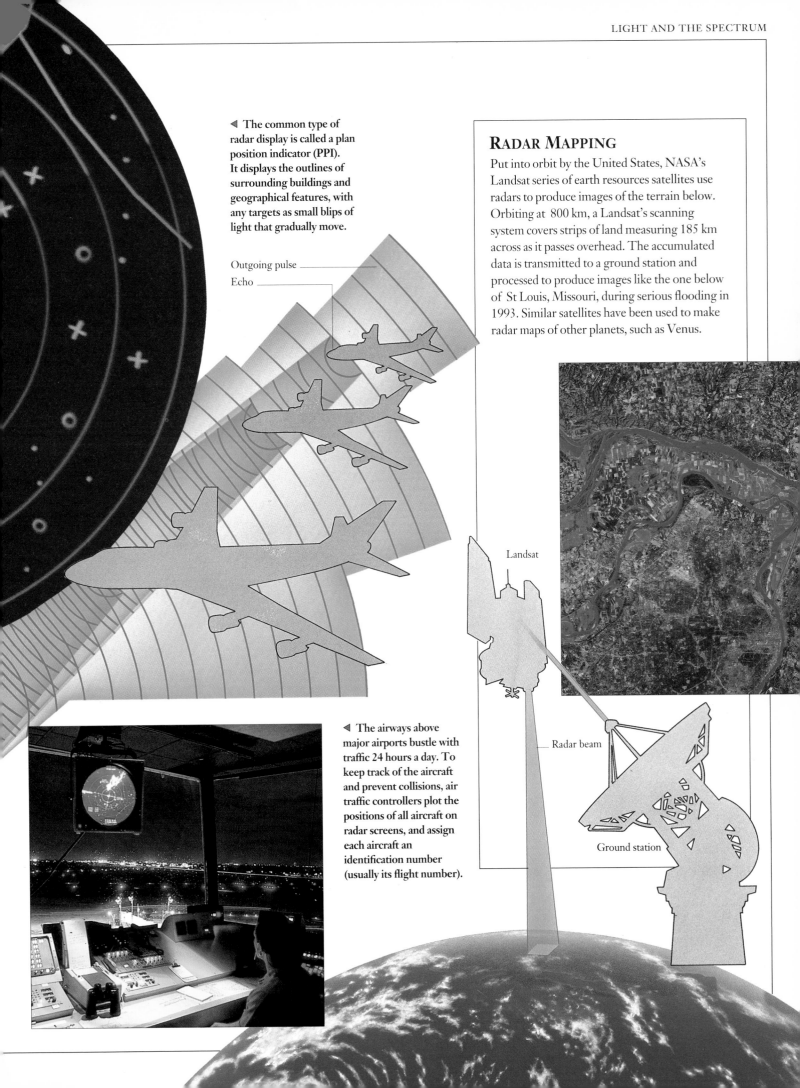

TV CAMERA AND RECORDER

TELEVISION is a means of sending pictures, usually by means of short-wavelength radio waves. The pictures may also be recorded on magnetic tape. In addition to entertainment, television has applications in industry, medicine and surveillance.

In a television camera, a lens system focuses an image of a scene onto a picture tube. In a color camera, the incoming light is split by filtering mirrors into its primary colors – red, green and blue – and each of these components passes to its own picture tube.

Inside the tube, a beam of electrons scans the image and generates a signal current that varies with the brightness of the image. Finally, in broadcast television, the amplified signal current (or combined currents for color) passes to a radio transmitter where it modulates a carrier wave, in the same way as in sound broadcasting. Because television uses microwaves, it can be transmitted only over line-of-sight distances. Television transmitters have very tall masts, to give maximum range. In closed-circuit television, the picture signal travels directly to the receiver along a coaxial or fiber optic cable.

The key part of a television camera is the picture tube. It has a cathode (negative electrode) producing a beam of electrons which scan the back of a layer of material that conducts electricity when light falls on it. This photoconductive layer is in contact with a transparent, conductive layer called the target plate which acts as the tube's anode (positive electrode). The anode backs onto a glass faceplate.

The electron beam scanning the photoconductive layer charges it up. Light from the camera lens falling onto the layer makes it lose charge in proportion to the brightness of the light. When the beam again scans those areas, it recharges them and causes a signal current to flow from the target plate. For this reason, the strength of the signal current varies in proportion to the amount of light.

The signal from a television camera consists of a varying electric current. It can be recorded on magnetic tape, the way a tape recorder captures sound, or computer data is recorded (stored) on a magnetic disk. To contain so much information, videotape must be wider than audio tape and the tape speeds are much higher. Tight packing is achieved by rotating the recording head at an angle to the direction of tape transport to record the signal as a series of diagonal bands. Portable camcorders use narrower tapes, but have a very limited capacity and give poorer picture quality. Pictures can be recorded digitally – on a magnetic disk as in computers, or a laser-scanned optical disk like those in compact disc players.

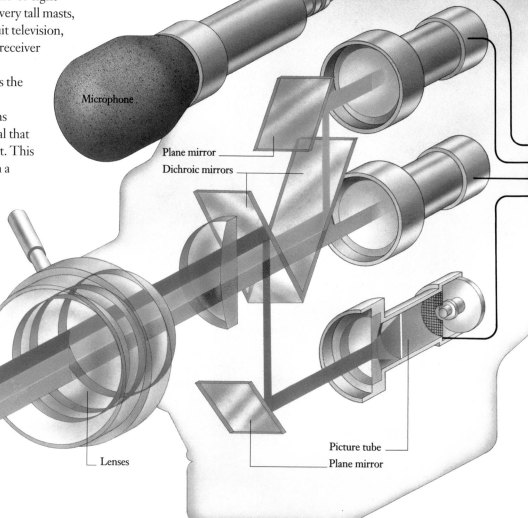

► Three main pieces of TV equipment are a color television camera, a video recorder and a portable camcorder (with a microphone to capture sound).

Microphone

Plane mirror

Dichroic mirrors

Lenses

Picture tube

Plane mirror

■ Many television news companies use fairly small portable TV cameras connected to video recorders BELOW, allowing greater mobility in covering the news. Alternatively, for "live" broadcast transmissions, the stationary TV cameras may be linked directly to a transmitter in a panel truck LEFT, which is parked nearby and beams signals back to the main TV studio control room.

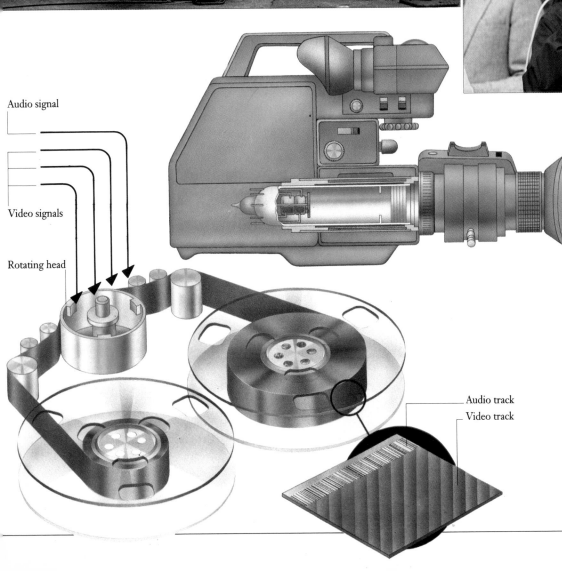

Audio signal

Video signals

Rotating head

Audio track

Video track

◄ In a color TV camera LEFT, the incoming light is split into three beams, one each for red, green and blue. Picture tubes convert each beam into electronic signals, which are mixed, along with luminance and chrominance information, to form a single video signal. Synchronization pulses and the audio signal are added, and the composite signal is stored on tape or sent to a trans-mitter for broadcasting. The recording head of a VCR CENTER rotates to pack signals on the tape as diagonal bands. Sound signals are recorded on the edge of the tape. A camcorder FAR LEFT is a small portable video camera that uses charged couple devices instead of picture tubes.

Television Playback

THE purpose of a television receiver is to take the broadcast signal picked up by a TV antenna and convert it into pictures and sound. In a color receiver, the incoming signal is first decoded – that is, split up to recreate the three signals used for color. The sound signal is also extracted at this stage, to be amplified and fed to a loudspeaker.

The color signals each pass to their own electron guns within the receiver's cathode-ray tube, one gun for each color. The resulting three electron beams pass between deflector coils which make them scan the inside of the face of the tube. A mask consisting of thousands of holes (in a showmask tube) or a vertical grille (in a Trinitron tube) guide the beams onto dots or stripes of red, blue and green phosphors. These fluoresce to produce red, blue or green light when struck by electrons, to produce a picture that gives the viewer the impression of full color.

The broadcast television pictures in most countries consist of 525 or 625 horizontal lines, built up at a rate of 25 or 30 complete pictures per second. The timer circuit, which controls the rate, works off the electricity mains at a frequency of 50 or 60 hertz (cycles per second), depending on the country in which the broadcast takes place.

At such picture speeds, the viewer's brain interprets the images as a continuous picture, not a series of individual ones. But if a television screen is photographed with a still camera using a shutter speed of faster than a 60th of a second, the photograph often shows only the part of the picture that was built up when the camera's shutter opened.

High definition television (HDTV), currently in development, uses twice as many scan lines and consequently gives pictures of much higher definition. Modern digital techniques allow other signals to be "tucked in" among the sound and picture signals. One example is Teletext, which is decoded by special circuits in the receiver and displayed on the screen when it is required.

Another recent development is flat-screen television, in which the screen is made up of thousands of tiny illuminated pixels. Small portable televisions, some tiny enough to be worn on the wrist like a watch, use screens with displays of liquid crystals (LCDs). Both pixels and LCDs change color when an electric field is applied to their edges.

A television receiver can also be driven by the output from a video recorder, which works in much the same way as a tape recorder on playback. Playing back the video tape results in a picture signal, which is fed into the receiver via its antenna socket. Signals picked up from a magnetic or nonmagnetic video disk can work a television receiver in a similar way.

Even broadcast television programs need not use radio waves. Alternative methods of distribution – particularly in densely populated areas – include cable television. This uses either microwave-carrying coaxial cables or, increasingly, optical fibers, which have the additional advantage of an extremely large bandwidth, allowing them to carry many channels simultaneously. Such cables can also carry computer data and other forms of digitized information.

KEYWORDS

CATHODE RAY TUBE
COLOR
ELECTRON
LIQUID CRYSTAL DISPLAY
MAGNETIC TAPE
PHOSPHOR
PIXEL
TELEVISION
VIDEO

TV signal

▲ Within line-of-sight of a local transmitter broadcasting on UHF frequencies, a horizontal or vertical "toast rack" antenna is used. Satellite transmissions need a dish antenna.

Picture Scanning

The lines scanned on the face of a TV picture tube are interlaced to give good quality pictures. First the "odd" lines (lines 1,3,5,7, and so on) are scanned from top to bottom. Then the scanning beam flips back to the top of the screen, and draws in the "even" lines (2,4,6, and so on). The whole sequence takes one twenty-fifth or one thirtieth of a second (depending on the mains frequency, which is 60 hertz – cycles per second – in North America but 50 hertz in Europe). Finally the beam flips back and repeats the whole process.

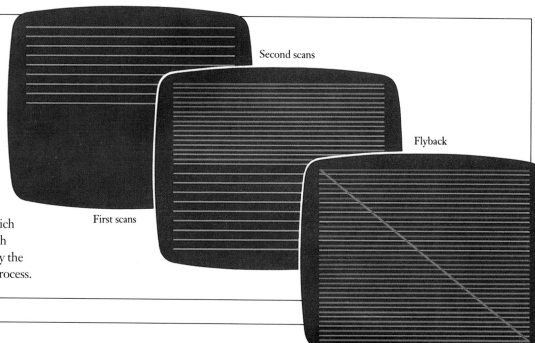

Second scans

First scans

Flyback

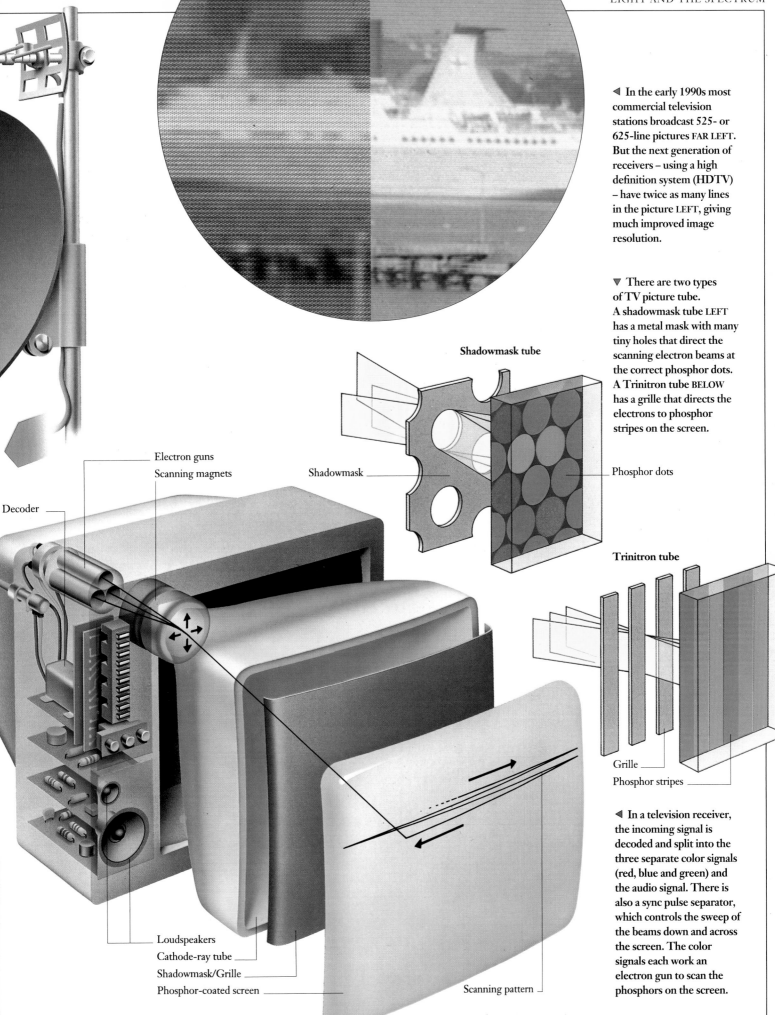

◀ In the early 1990s most commercial television stations broadcast 525- or 625-line pictures FAR LEFT. But the next generation of receivers – using a high definition system (HDTV) – have twice as many lines in the picture LEFT, giving much improved image resolution.

▼ There are two types of TV picture tube. A shadowmask tube LEFT has a metal mask with many tiny holes that direct the scanning electron beams at the correct phosphor dots. A Trinitron tube BELOW has a grille that directs the electrons to phosphor stripes on the screen.

Shadowmask tube

Shadowmask

Phosphor dots

Trinitron tube

Grille

Phosphor stripes

Electron guns

Scanning magnets

Decoder

Loudspeakers

Cathode-ray tube

Shadowmask/Grille

Phosphor-coated screen

Scanning pattern

◀ In a television receiver, the incoming signal is decoded and split into the three separate color signals (red, blue and green) and the audio signal. There is also a sync pulse separator, which controls the sweep of the beams down and across the screen. The color signals each work an electron gun to scan the phosphors on the screen.

6

INSIDE
the Atom

ANY PHYSICAL PHENOMENA can be explained in terms of the behavior of atoms. Most optical and electrical effects, for example, involve the production or movement of electrons – negatively charged subatomic particles in orbit around a central, positively charged nucleus. The orbits correspond to specific energy levels, and the movement of electrons between these levels produces light.

Another group of phenomena result from changes within the nucleus itself. The spontaneous disintegration of the nucleus in certain atoms produces radioactivity – the emission of energy in the form of radiation, accompanied by changes in the identity of the atoms themselves.

The nucleus consists of a collection of two kinds of subatomic particles: positively-charged protons (whose charges balance those of the electrons in an uncharged atom) and uncharged neutrons. Protons and neutrons have similar masses, and both are about 1800 times as massive as an electron. The simplest chemical element, hydrogen, is an exception: its nucleus has no neutrons.

According to modern atomic theory, there are also many other subatomic particles. Even the proton and neutron are thought to consist of smaller components called quarks.

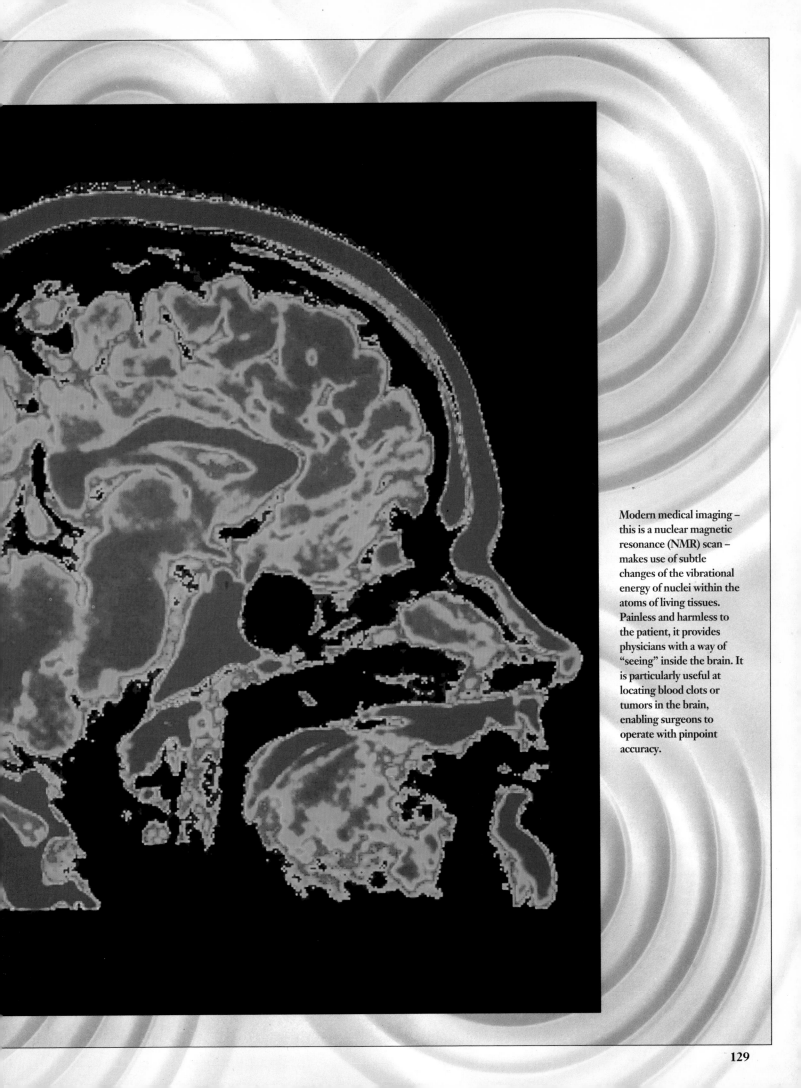

Modern medical imaging – this is a nuclear magnetic resonance (NMR) scan – makes use of subtle changes of the vibrational energy of nuclei within the atoms of living tissues. Painless and harmless to the patient, it provides physicians with a way of "seeing" inside the brain. It is particularly useful at locating blood clots or tumors in the brain, enabling surgeons to operate with pinpoint accuracy.

SUBATOMIC PARTICLES

THE principal subatomic particles – components of an atom – are the electron, proton and neutron. Protons and neutrons are normally located in an atom's nucleus, which is surrounded by orbiting electrons. A simple model of the atom gradually emerged in the early years of the 20th century as the various particles were discovered.

The electron was discovered by the British physicist Joseph Thomson (1856–1940) in 1897.

He studied cathode rays, emitted by the cathode (negative electrode) of a vacuum tube. Thomson showed that the "rays" actually consist of tiny charged particles, which he called electrons. The term "cathode ray" persists in the name of the cathode-ray tube used in television sets, radar displays and computer terminals. All of these devices utilize a beam of electrons (from a cathode) in order to excite the light-emitting phosphors on the inside of the tube's screen.

The existence of the tiny atomic nucleus, which carries most of the atom's mass but not its electrons, was deduced from experiments carried out by the New Zealand-born physicist Ernest Rutherford (1871–1937) in 1911. He demonstrated that the nucleus has a positive charge, and later discovered that

▶ **In an experiment carried out in 1909 by Ernest Rutherford's assistant Ernest Marsden, a beam of alpha rays (positive particles emitted by radium) was fired at a thin sheet of gold foil. Most of the particles passed straight through, but others were scattered through various angles. After considering these puzzling results, Rutherford concluded that they could be explained only if the gold atoms contained a central, positively-charged nucleus.**

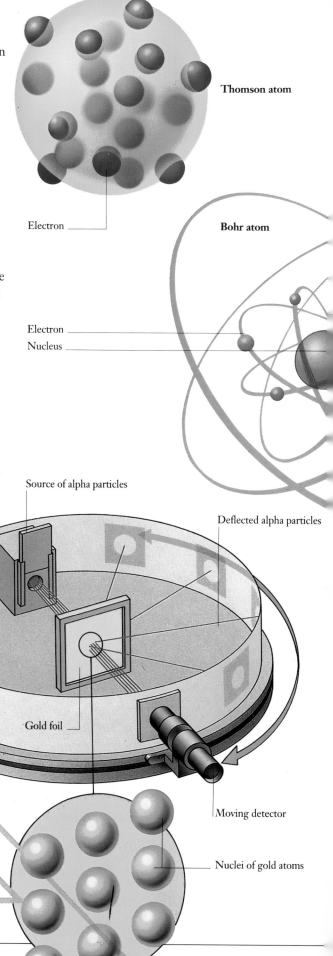

Thomson atom

Electron

Bohr atom

Electron
Nucleus

Source of alpha particles

Deflected alpha particles

Gold foil

Moving detector

Nuclei of gold atoms

Alpha particle

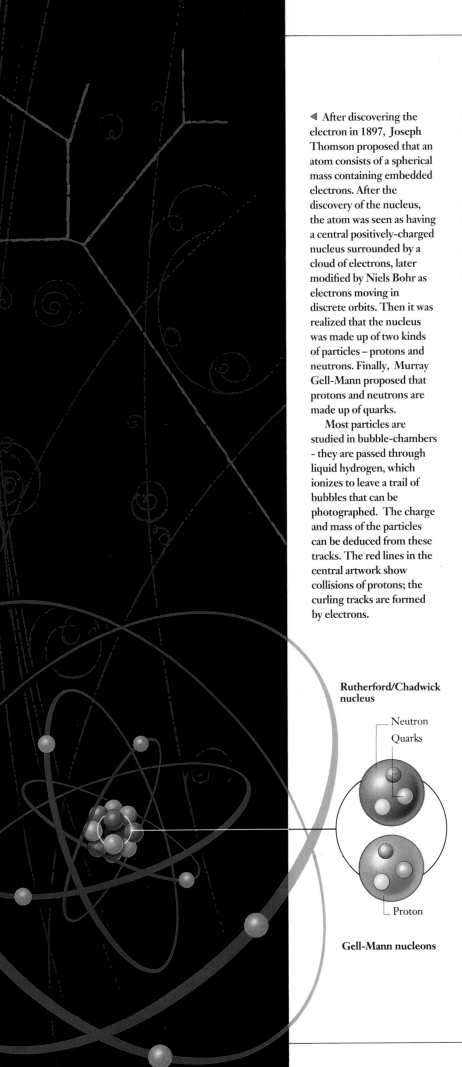

◄ After discovering the electron in 1897, Joseph Thomson proposed that an atom consists of a spherical mass containing embedded electrons. After the discovery of the nucleus, the atom was seen as having a central positively-charged nucleus surrounded by a cloud of electrons, later modified by Niels Bohr as electrons moving in discrete orbits. Then it was realized that the nucleus was made up of two kinds of particles – protons and neutrons. Finally, Murray Gell-Mann proposed that protons and neutrons are made up of quarks.

Most particles are studied in bubble-chambers - they are passed through liquid hydrogen, which ionizes to leave a trail of bubbles that can be photographed. The charge and mass of the particles can be deduced from these tracks. The red lines in the central artwork show collisions of protons; the curling tracks are formed by electrons.

Rutherford/Chadwick nucleus

Neutron
Quarks

Proton

Gell-Mann nucleons

the charge carrier is the proton (whose charge is equal but opposite to that of an electron). Rutherford also showed that a proton is very much more massive than an electron.

The number of protons in an atom's nucleus equals the number of orbiting electrons, so that the atom has no overall electric charge. Every element has a different number of protons, and it is this number – the atomic number – that gives a particular element its identity. But the protons alone do not account for the total mass of a nucleus (except in the case of hydrogen). The remainder of the nuclear mass comes from neutrons, the other major subatomic particle, discovered in 1932 by the British physicist James Chadwick (1891–1974). A neutron carries no electric charge, and has a mass almost exactly the same as that of a proton.

Later discoveries made this simple model of the atom – protons and neutrons in the nucleus, surrounded by electrons in orbit – obsolete. In addition to protons, neutrons and electrons, other subatomic particles began to turn up, often after their existence had been predicted to account for the behavior of atoms and other particles. These included the pi meson (with mass between that of a proton and an electron) and the positron (a positively-charged particle with mass equal to that of an electron). Since the development in the 1930s of particle accelerators such as the cyclotron, collectively nicknamed "atom smashers", more than 30 other particles have been identified, although their role in atomic – usually nuclear – structure was not always clear. They are classified into two main types: those with no apparent internal structure, such as the electron, positron, muon, neutrino and tau particle, collectively called leptons; and the group known as hadrons – proton, neutron, pi meson – which do have an internal structure.

Another particle, the quark, was proposed in 1962 by the United States physicist Murray Gell-Mann to clarify the confusion caused by the incomplete identification of subatomic particles. The quark is a hypothetical fundamental particle devised to explain the existence and behavior of all the others. At least 12 different kinds of quark (and antiquark) have now been described. They combine to form protons, neutrons, and various other subatomic – but no longer fundamental – particles. Quarks have fractional electric charges of either $+\frac{2}{3}$ or $-\frac{1}{3}$. A proton consists of three quarks of charges $\frac{2}{3}$, $\frac{2}{3}$ and $-\frac{1}{3}$, giving an overall charge of +1. The three quarks in a neutron have charges of $\frac{2}{3}$, $-\frac{1}{3}$ and $-\frac{1}{3}$, resulting in an overall charge of zero. Pi mesons consist of only two quarks.

THE UNSTABLE ATOM

AN ELEMENT'S chemical identity is determined by the number of protons in the nuclei of its atoms. Elements are listed in the Periodic Table in order of the number of protons in the nucleus – their atomic number. Hydrogen has 1 proton, helium 2, lithium 3, beryllium 4, and so on right through to the heaviest naturally occurring element, uranium, with 92 protons. There are a dozen or so even heavier elements, but these have all been made in the laboratory and do not occur naturally on Earth.

The number of neutrons in the atomic nuclei also increases along the list, but in a less predictable way. Helium has 2 neutrons, lithium 3, beryllium 5 and uranium as many as 146. Uranium and some other elements have two or more forms with different numbers of neutrons in their nuclei. These forms, called isotopes, have the same numbers of protons (and hence the same chemical identity), but the differing number of neutrons gives them different masses. Uranium, for example, has isotopes of masses 234, 235 and 238.

Some combinations of protons and neutrons are unstable because of forces acting within their nuclei. They spontaneously disintegrate, usually by emitting an alpha particle (two protons and two neutrons) or a beta particle (an electron). Penetrating gamma rays may be emitted at the same time, and the whole phenomenon is called radioactivity. Because an alpha emitter loses two protons and two neutrons, it turns into a different element whose atomic number is therefore two less. Radium-226 (the isotope of radium with mass 226), of atomic number 88, decays by emitting alpha particles to become the gaseous element radon, atomic number 86, mass 222.

In the nucleus of a beta-emitting radioactive isotope, a neutron changes into a proton and releases an electron (beta particle). This results in no significant change in mass but an increase of one in atomic number, with again the formation of a different element. For example, radium-228 (atomic number 88) emits beta particles to become actinium-228 (atomic number 89). In both of these examples, radioactive decay takes the form of a cascade, resulting in products that are themselves radioactive, and go on to form yet other elements – which in turn experience radioactive decay. Uranium-238 decays by emitting both alpha and beta particles, passing

KEYWORDS

ALPHA PARTICLE
ATOMIC NUMBER
BETA PARTICLE
DECAY
GAMMA RADIATION
HALF-LIFE
ISOTOPE
RADIOACTIVITY
RELATIVE ATOMIC MASS

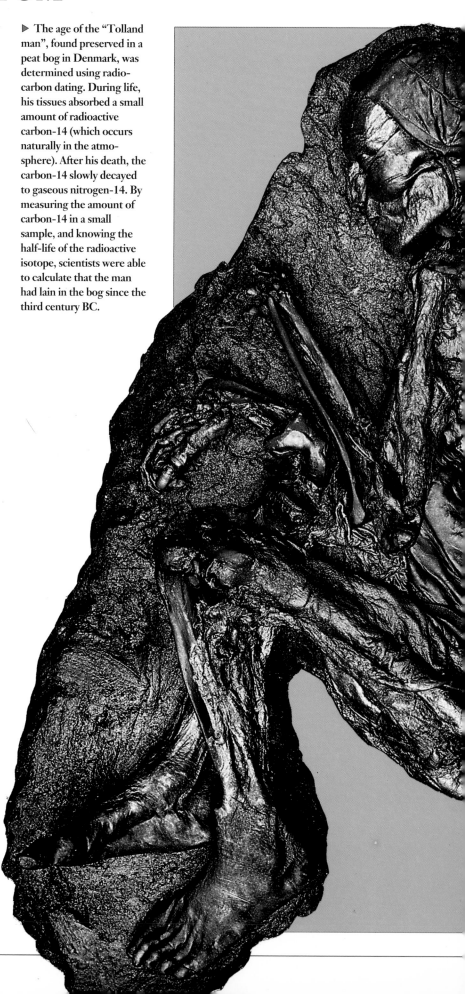

▶ The age of the "Tolland man", found preserved in a peat bog in Denmark, was determined using radio-carbon dating. During life, his tissues absorbed a small amount of radioactive carbon-14 (which occurs naturally in the atmosphere). After his death, the carbon-14 slowly decayed to gaseous nitrogen-14. By measuring the amount of carbon-14 in a small sample, and knowing the half-life of the radioactive isotope, scientists were able to calculate that the man had lain in the bog since the third century BC.

through 14 different stages before eventually finishing as the stable element lead-206.

Not all radioactive isotopes decay at the same rate; nor do all the atoms in a sample disintegrate simultaneously. Physicists express the rate in terms of an element's half-life – the time it takes for half the atoms to decay into other atoms. The naturally occurring element thorium-232 has a half-life of nearly 14 billion years; some isotopes, created in cyclotrons, have half-lives of only a few hundred thousandths of a second.

Identifying a radioactive isotope in an object, and knowing its half-life, provides a way of estimating the object's age. Anything made of wood contains tiny amounts of radioactive carbon-14, absorbed from the atmosphere while the tree was alive. In radiocarbon dating, the amount of carbon-14 is measured and the age of the wood calculated from the knowledge that carbon-14's half-life is 5770 years (it decays to form ordinary nitrogen-14). Radiocarbon dating can also be used to estimate the age of bone, leather, paper or even hair – in fact anything that contained carbon during its life. A similar method, potassium-argon dating, is used to determine the ages of rocks.

▼ **In a search for new particles, a scientist checks part of the detector on a collider at the 27-kilometer accelerator at CERN, near Geneva, Switzerland. The accelerator speeds electrons and positrons using 50 billion electron** volts, and the detector studies what happens when the particles – traveling in opposite directions – are made to collide. Shortlived subatomic particles may result, enabling scientists to study the structure and origins of matter.

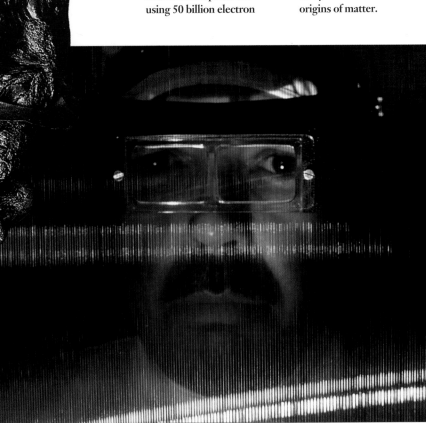

NATURAL DECAY

There are three natural radioactive decay series, beginning with the isotopes thorium-232, uranium-235 and uranium-238 (shown here). This series involves a chain of 14 different radioactive isotopes – of thorium (Th), protactinium (Pa), radium (Ra), radon (Rn), polonium (Po), lead (Pb) and bismuth (Bi) – before ending with the nonradioactive lead-206. Each isotope in the chain emits alpha particles, beta particles, gamma rays or a combination.

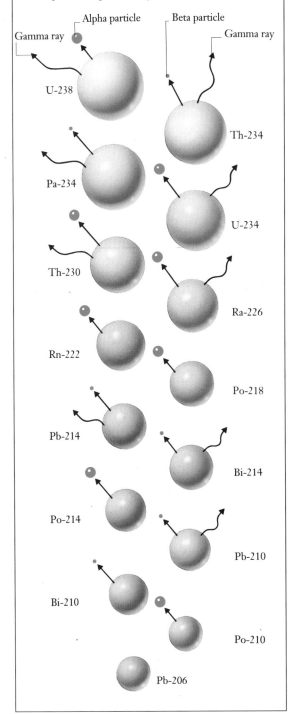

NUCLEAR FISSION

S LOW spontaneous disintegration in the nuclei of their atoms makes radioactive elements unstable. Bombarding the nuclei with neutrons speeds up the disintegration or makes a stable nucleus unstable. The nuclei absorb the neutrons and split, releasing energy. This is because the mass of the fission products is fractionally less than the mass of the split nucleus. This "lost" mass is converted into energy.

Nuclear fission (splitting the nucleus) sometimes produces more neutrons. If these in turn are absorbed by other nuclei which then split, producing even more neutrons, a rapidly accelerating process called a chain reaction may result. Uncontrolled, it results in a nuclear explosion. A controlled reaction forms the basis of a reaction that can be harnessed as energy.

One of the first materials used in controlled fission – in a nuclear reactor – was the isotope uranium-235. It occurs as a very small percentage (0.72 percent) of natural uranium, which consists mainly of stable uranium-238. When removed from natural uranium, uranium-235 can be used as a nuclear fuel.

Fission is more likely to occur if the bombarding neutrons are moving slowly. The uranium fuel in a reactor is surrounded by a moderator such as graphite or heavy water – deuterium oxide (D_2O). The reactor also has control rods which can be moved into or out of the reactor core to speed up or slow down the chain reaction by absorbing some of the bombarding

■ A scanning transmission electron micrograph RIGHT shows hexagonally packed uranium atoms in a crystal of an organic uranium salt. Uranium-235 was one of the first fissionable isotopes used in nuclear weapons and reactors. The sequence BELOW shows how the chain reaction – causing fission – takes place. An incoming slow, or thermal, neutron combines with a U-235 nucleus to form unstable U-236 which immediately splits to form two smaller atoms (such as barium and krypton) with the release of three neutrons. These neutrons go on to split three more U-235 nuclei, releasing nine neutrons, and so on as the chain reaction rapidly accelerates.

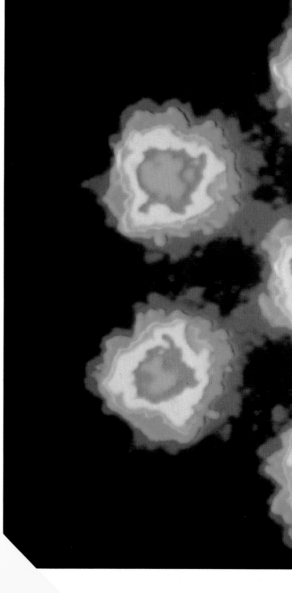

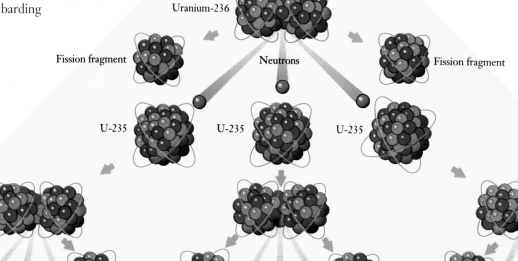

Slow neutron

Uranium-235

Uranium-236

Fission fragment   Neutrons   Fission fragment

U-235   U-235   U-235

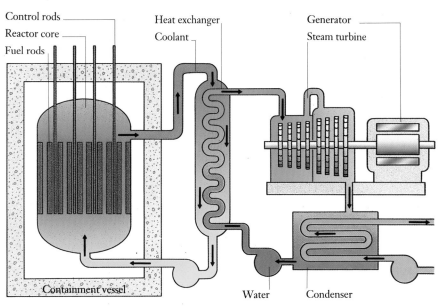

Control rods
Reactor core
Fuel rods
Heat exchanger
Coolant
Generator
Steam turbine
Containment vessel
Water
Condenser

■ At a nuclear power plant ABOVE, the reactor is merely a type of boiler for heating water to make steam. A gas or liquid coolant (which may be pressurized water, liquid sodium) flowing through the hot reactor core passes beyond the heavily shielded containment vessel to enter a heat exchanger. There it boils water and the steam passes to steam turbines connected to an electricity generator. Smaller reactors have been used in submarines and ships, such as the icebreaker LEFT.

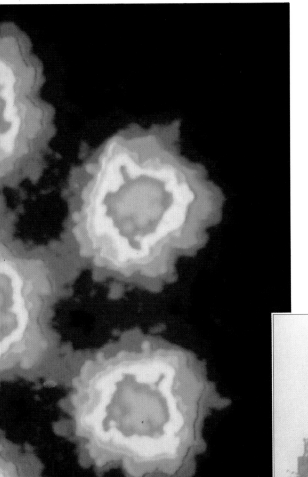

▶ Nearly all the world's nuclear reactors are used for generating electricity, particularly in countries (such as France) with poor reserves of fossil fuels. The difficulty of decommissioning reactors and processing or storing spent nuclear fuel, has made them less popular than they once were. This image shows the gantry above the reactor core, from which control and fuel rods can be lowered.

neutrons. The elements boron and cadmium make efficient control rods.

The intense heat produced inside the reactor's core is absorbed by a fluid and carried to an external heat exchanger. The fluid becomes highly radioactive and is recycled around a closed system within the reactor. Most reactors use water as a coolant kept under pressure to let it reach a high temperature without boiling (a pressurized-water reactor). An advanced gas-cooled reactor uses carbon dioxide gas instead of water. The heat carried away to the heat exchanger is used to boil water for making steam, which drives turbines for generating electricity.

A design using enriched uranium or plutonium-239 as fuel is called a fast reactor because no moderator is employed, the core gets so hot that liquid sodium metal has to be used as a coolant.

NUCLEAR FUSION

FUSION is a nuclear reaction in which the nuclei of light atoms combine to form heavier, more stable nuclei. It takes place in the Sun and other stars, which "burn" hydrogen in their cores. Two hydrogen nuclei (protons) combine to form deuterium, which is an isotope of hydrogen with a mass of 2. Further fusion of the deuterium results in the formation of helium, with the release of vast amounts of heat and light. As in nuclear fission, the mass of the products is slightly less than those of the reactants, and the missing mass appears as energy. However, unlike fission, the products of fusion are not radioactive. This makes fusion reactors attractive as a safe power source, along with low cost: deuterium can be easily obtained from sea water, and tritium can be made from the fairly common element lithium.

Scientists trying to copy this reaction on Earth discovered that the best starting materials are deuterium and tritium (another isotope of hydrogen with a mass of 3). In the basic fusion reaction, an atom of deuterium combines with an atom of tritium to form an atom of helium, with the release of a neutron and vast amounts of heat energy.

Because of the similar positive electric charges on their nuclei, the isotopes are reluctant to fuse except at extremely high temperatures (about 100 million degrees). Such temperatures were first achieved artificially in a hydrogen bomb, in which the fusion reaction is allowed to go out of control.

Today scientists are trying to produce controlled fusion. The chief difficulty is that, at such high temperatures, the combining gases exist as the fourth state of matter known as a plasma. The negative electrons and positive hydrogen nuclei are separated to give a completely ionized fluid, which can be contained only by powerful magnetic fields – no physical container could withstand the extremely high temperatures produced in a fusion reaction.

Attempted solutions to the fusion problem usually involve containing the plasma in a magnetic "bottle", which may be shaped like a figure of eight or like a doughnut (called a torus). The Tokamak reactor consists of a torus surrounded by D-shaped coils. Pulses of high-voltage current in the coils create the plasma and raise its temperature, and the powerful magnetic field produced constrains the plasma to a spiral path around the middle of the torus.

■ The Sun RIGHT and other stars derive their vast amounts of energy from fusion reactions taking place in their interiors. Chief of those in the Sun involve hydrogen (BELOW RIGHT, UPPER BAND). First, hydrogen atoms combine to form the two-atom hydrogen isotope deuterium. The deuterium atoms then combine in a fusion reaction to form helium, with the release of energy and two neutrons. Attempts to produce controlled fusion on Earth also use deuterium, but the other component of the reaction is the three-atom hydrogen isotope called tritium (BELOW RIGHT, LOWER BAND). Again the product of fusion is an atom of helium and a single neutron.

▲ A Tokamak fusion reactor has a donut-shaped "bottle" containing a plasma of deuterium and tritium. Powerful D-shaped magnets sit inside a huge transformed core. The magnets create a field which, in combination with that of the current passed through the plasma, produces helical field lines around the high-temperature plasma and that keeps it away from the walls of the vessel. The design has proved to work, but the prospect of producing power economically from fusion remain many decades in the future.

Plasma
Vacuum vessel
Field coil

▼ **Two establishments are building Tokamak reactors: Princeton University in the United States (shown here) and the Joint European Torus (JET) project near Oxford, England. Both designs are similar and their function is to provide heat to boil water; the steam would work turbines to power electricity generators.**

An alternative method uses a long magnetic "bottle", sealed at each end by magnetic mirrors. Tiny glass pellets with fusible material (deuterium and tritium gas) are introduced into the bottle and bombarded by flashes of laser light. This vaporizes the glass and raises the gases to the plasma state and fusion temperature.

Research continues into magnetic confinement and laser implosion, but a sustained, controlled fusion reaction has yet to be demonstrated. Temperatures above 100 million degrees have been attained in Tokamak experiments, but not at the same time as the required pulsating magnetic field.

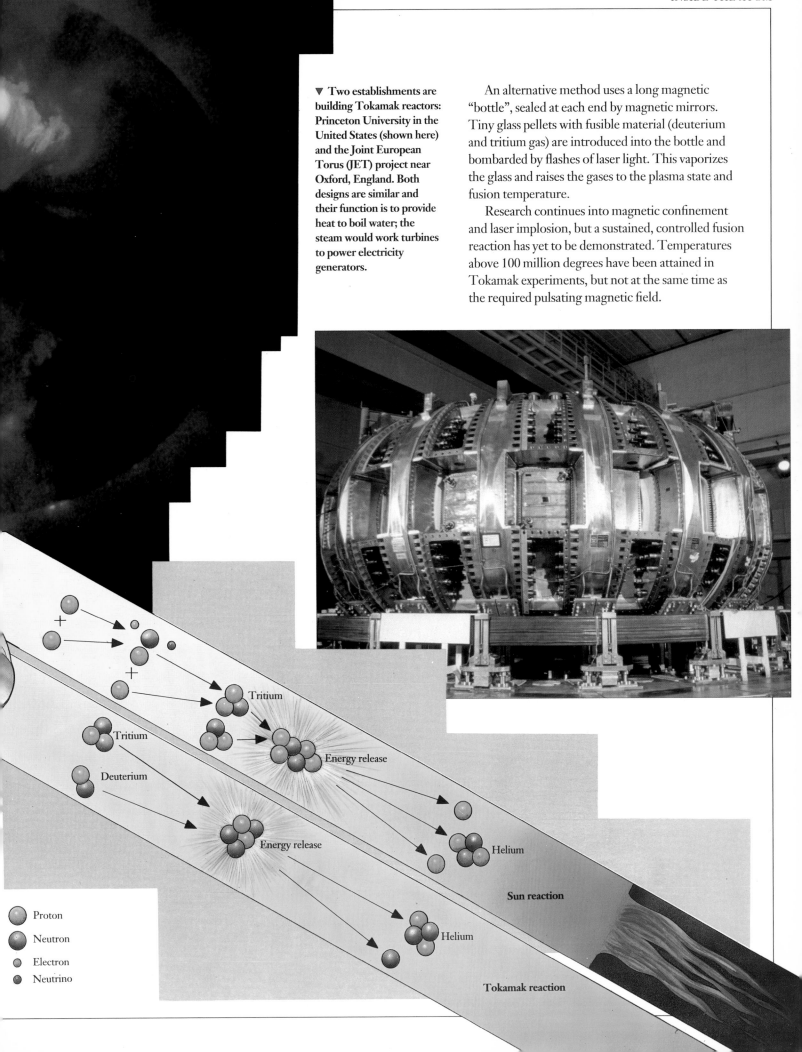

Tritium

Tritium

Deuterium

Energy release

Energy release

Helium

Sun reaction

Helium

Tokamak reaction

Proton

Neutron

Electron

Neutrino

QUANTUM PHYSICS

WHEN atoms emit electromagnetic radiation – as when atoms of a metal heated to the point of incandescence give off light – the radiation has a characteristic wavelength rather than a continuous range of energies. Light is emitted when electrons "jump" between an orbit of high energy to one at a lower energy level. Each jump releases a "packet" of light energy, called a quantum, equal to the difference in energies between the two atomic orbits. Light quanta are known as photons, and in many respects they can be regarded as particles with a mass of zero but a specific amount of energy and momentum.

The particle theory of light has potential practical applications. One proposal for powering future inter-stellar spacecraft uses huge sails to gather sunlight. The millions of photons hitting the sails would push the craft along. The principle is easy to understand in terms of particle-like photons; it is more difficult if light is thought of as electromagnetic waves.

Another phenomenon that could not be explained in terms of the wave theory of light is the photoelectric effect. This is the release of electrons from the surface of a material (usually a metal) when it is struck by light, X rays or gamma rays. Atoms in the material absorb photons from the incident radiation. The photons transfer sufficient energy to some electrons to enable them to escape.

Many other physical phenomena described in terms of waves of energy can also be explained by quantum theory. But there is no one "correct" theory: waves or quanta can be chosen, whichever is convenient for the purposes of explanation.

In fact, electrons in an atom do not actually orbit the nucleus in fixed circular orbits like planets orbiting the Sun. A better model, developed alongside quantum theory, visualizes

▶ **Proof of the quantum nature of light is the photoelectric effect, in which the number and energy of electrons emitted from a metal struck by light depends the intensity and frequency of the light.**

Electron

Low frequency, low
energy photons

Low frequency, high
energy photons

High frequency, low
energy photons

High frequency,
high energy photons

electrons as occupying regions in space, and there is a definite probability that a particular electron will be in a certain place at a certain time. The region in which the electron is likely to be found is called an orbital.

Probability – not certainty – is involved because of a principle put forward by the German physicist Werner Heisenberg in 1927. Heisenberg argued that it is not possible to know simultaneously the precise position and momentum of an electron. Probability is a mathematical concept, and modern physicists and chemists deal with electrons using equations whose terms are probability functions. This approach is known as quantum mechanics.

Physicists now deal with many other kinds of quanta, as well as photons. For example, the thermal vibrations of atoms in the lattice of a crystal correspond to discrete energy states whose quantum is the phonon. A roton is a quantum of rotational energy, and a magnon is the quantum of spin energy of the molecular magnets in a magnetic substance.

Electrons in an atom and the nuclei themselves also spin, and this spin is quantized (and ascribed a quantum number). Transitions or "jumps" between the two possible spin states of the outer electron in a cesium atom are exploited in the cesium clock, which is accurate to within a second in a million years.

◀ A long-exposure photograph of moving water exemplifies Heisenberg's uncertainty principle: it is impossible to know simultaneously the exact position and momentum of a subatomic particle in motion. In order to determine its position, the particle has to be illumi-nated (to be "seen"). But photons of the illuminating radiation will strike the particle and change its momentum, moving it from its previous position.

▶ A cesium clock is the most accurate timepiece ever made, keeping time with an incredible accuracy of one second in a million years. It makes use of a quantum effect that occurs in atoms of the rare metal cesium. Its nucleus can exist in either of two energy states, and supplying cesium atoms with microwave energy excites some nuclei to the higher energy state. These are then used to control the clock.

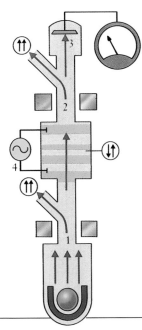

◀ In a cesium atomic clock, the natural vibrational frequency of some cesium electrons is used to "tune" the output of a microwave oscillator. When cesium ions 1 – produced by heating some of the metal – are irradiated with microwaves, some of their electrons increase in energy and are deflected by a magnetic field 2 to collide with a detector 3. A signal from this detector is fed back to the microwave generator 4 to lock it onto the correct frequency (which is equal to 9,192,631,770 hertz).

WAVES AND PARTICLES

ELECTRONS were discovered first in cathode rays, which are streams of particles given off by a cathode in a vacuum tube (as in a cathode-ray tube). Physicists soon determined the mass of the electron, and found it to be about 0.0005 of that of a hydrogen atom. Today they use sophisticated equipment to accelerate electrons to very high speeds and use them as "bullets" to smash atoms in cyclotrons.

Light waves can be considered as made up of particles (photons). If a light wave can be made up of particles, could a stream of electrons – which are usually regarded as tiny particles – behave like a wave? This question was first asked by the French physicist Louis de Broglie, who in 1924 proposed that the answer is yes.

One phenomenon of light that demonstrates its wave nature is interference – the alternate reinforcement and cancellation that take place when two similar waves come together, creating light and dark bands or rings. A few years after de Broglie's prediction, scientists in the United States and Britain produced interference patterns by scattering electrons from the surface of a metal crystal or in thin metal foil. They had demonstrated the existence of electron waves.

Another well-known property of light is that it can be focused by lenses to make optical instruments such as telescopes and microscopes. Because electrons carry a (negative) electric charge, a beam of electrons can be bent by a magnetic field. A circular electromagnet around an electron beam can be used to focus it just as a lens focuses light. A series of magnetic lenses are employed in this way in an electron microscope, which utilizes the very short wavelength of electron waves to produce highly magnified images of objects that are far too small to be seen even with the most powerful optical microscopes.

These discoveries gave new meaning to quantum theory. The probability of locating an electron in an atom turns out to be a wave function that describes the state of an electron in a given orbital, including such factors as its spin, angular momentum and likely position in space. For example, an atom's innermost – lowest energy – electrons have zero angular momentum and are located in a spherical orbital centered on the nucleus. The electrons in the next highest energy level have an angular momentum of 1

KEYWORDS

ANGULAR MOMENTUM
ELECTRON
ELECTRON MICROSCOPE
INCANDESCENCE
LASER
ORBITAL
PARTICLE ACCELERATOR
PHOTON
QUANTUM THEORY
SPECTRUM
SUPERCONDUCTIVITY
WAVE MECHANICS

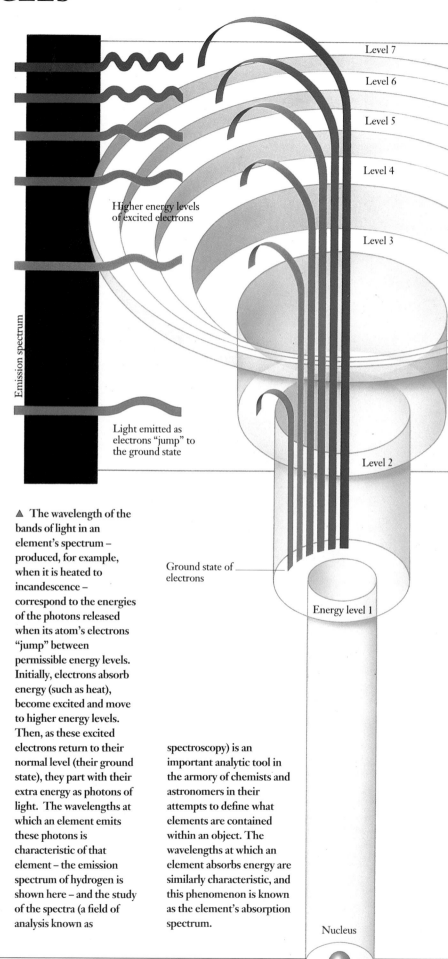

▲ The wavelength of the bands of light in an element's spectrum – produced, for example, when it is heated to incandescence – correspond to the energies of the photons released when its atom's electrons "jump" between permissible energy levels. Initially, electrons absorb energy (such as heat), become excited and move to higher energy levels. Then, as these excited electrons return to their normal level (their ground state), they part with their extra energy as photons of light. The wavelengths at which an element emits these photons is characteristic of that element – the emission spectrum of hydrogen is shown here – and the study of the spectra (a field of analysis known as spectroscopy) is an important analytic tool in the armory of chemists and astronomers in their attempts to define what elements are contained within an object. The wavelengths at which an element absorbs energy are similarly characteristic, and this phenomenon is known as the element's absorption spectrum.

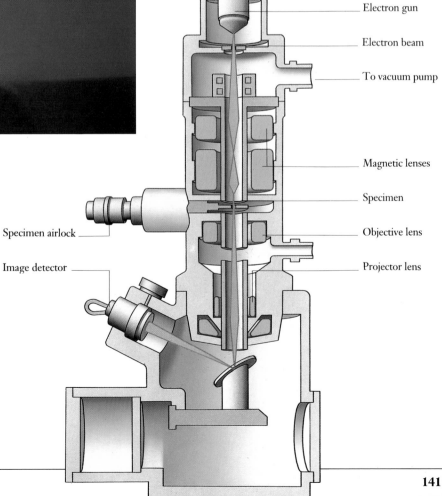

◀ Electron microscopes produce images of objects too small to be resolved by even the most powerful optical microscopes. Alternatively, using scanning techniques, they can record detail of larger objects, such as a fly LEFT. To be "seen" by the microscope, the sample is coated in metal FAR LEFT.

▼ The resolution of an optical microscope is limited by the wavelength of light (about 10^{-7} meters). To view objects smaller than this, scientists employ electron microscopes, in which beams of electrons behave as radiation of extremely short wavelength (down to 10^{-15} meters or less). In an optical microscope, the light rays are focused by giant lenses.

But because electrons are negatively charged particles, they are focused in an electron microscope by circular magnets. The inside of the apparatus, including the sample, is kept in a high vacuum. The image may show a section of the subject (a transmission micrograph) or a three-dimensional representation (a scanning electron micrograph).

Electron gun

Electron beam

To vacuum pump

Magnetic lenses

Specimen

Specimen airlock

Objective lens

Image detector

Projector lens

and occupy three dumbbell-shaped orbitals aligned at right angles to each other. In this way, waves are equated with probabilities, which in turn specify shapes.

Using this combined theory, called wave mechanics, it is now possible to explain all the phenomena of physics, from the multicolored spectra of incandescent atoms and the production of laser light to the superconducting behavior of metals at temperatures near absolute zero. In particular, the characteristic light emitted by a particular element – its spectral "fingerprint" – corresponds to a wave, or photons, emitted as electrons change orbitals after absorbing energy. This explanation was first offered as long ago as the year 1900 by the German physicist Max Planck when he first proposed quantum theory, which was to have such a dramatic effect on 20th-century physics.

MASS EQUALS ENERGY

LARGE amounts of energy – mostly heat – are produced by nuclear fission or fusion because the masses of the products of the reaction are slightly less than the masses of the reactants. This "lost" mass – the difference between the two – appears as energy. The interchangeability of mass and energy was predicted in 1905 by the German-born physicist Albert Einstein in his special theory of relativity.

A key principle behind the theory is that the speed of light in a vacuum is constant, even if the source of light is moving in relation to an observer. If light is projected from the aircraft, it moves forward at a constant speed no matter how fast the plane is flying. The speed of light is nearly 300,000 kilometers per second, and is represented by the letter c.

KEYWORDS

ENERGY
FUNDAMENTAL FORCES
LIGHT
MASS
SPEED OF LIGHT
RELATIVITY

Strange effects occur when something moves at very high speeds. To a stationary observer, a hypothetical spacecraft traveling at half the speed of light appears to get shorter (by about 13 percent) and more massive (by the same proportion). The relative increase in mass equals the energy given to the spacecraft to accelerate it. Einstein showed that any mass has an energy equivalent, and that the energy is equal to the mass multiplied by the square of the speed of light. In mathematical terms, this is expressed by the equation $E = mc^2$.

Einstein adopted the accepted principle that light travels by the shortest distance between two points. But when he proposed that time is the fourth dimension, he found that massive objects in space, such as stars and galaxies, distort spacetime through gravitational effects. And where spacetime is curved, light follows a curved – though still the shortest – path.

▶ The equivalence of mass and energy, another result of relativity, is given an awesome proof by the explosion of a thermonuclear weapon. Such devices require pressure and temperature equivalent to those at the core of the Sun, fusing hydrogen nuclei into helium.

▼ A result of the special theory of relativity is that, as an object's speed approaches the speed of light, its mass increases exponentially and its length decreases to zero. For these reasons scientists argue that it is impossible for any object to travel faster than the speed of light.

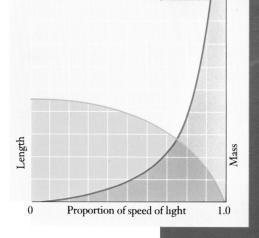

Length · Mass · 0 · Proportion of speed of light · 1.0

▶ If a railroad car has a bright light in the center, and devices that open the end doors when the light is turned on, people in the car will see both doors opening at exactly the same time 1. But to an observer standing by the track as the train speeds past, the back door opens before the front one. The back door travels forward to meet the approaching light, while the front door moves away from it 2.

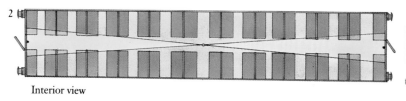

Interior view

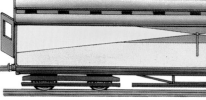

Exterior view

This behavior of light is a consequence of Einstein's general theory of relativity (1916). It has since been verified by various astronomical observations. In one of these, the path of light from a distant star is seen to be curved when it passes close to the Sun. More recently, radio astronomers have captured images of what appear to be a pair of quasars (very distant, very bright stars), located at the same distance from Earth. There is only one quasar, but radiation from it passes close to an intervening galaxy. The galaxy acts as a gravitational lens, bending the radiation so that, from Earth, it appears to originate in a different place from the radiation that travels directly to Earth.

Einstein also tried to combine his relativity theory with the quantum theory. This ideal, the unified field theory, still occupies many physicists, although others believe that it will never be achieved.

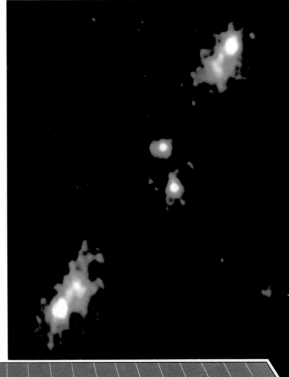

Quasar   Galaxy ⌐ ⌐ Bent light path   Quasar

▲ **Light reaching us from a distant source in space may travel close to a massive body, such as a galaxy. The general theory of relativity says that the gravity of the intervening body will bend** the path of the light, so that two versions of the distant object may be visible, as in the photograph ABOVE. A similar object viewed without an intervening lens would appear normal.

FACTFILE

PRECISE MEASUREMENT is at the heart of all science, and several standard systems have been in use in the present century in different societies. Today, the SI system of units is universally used by scientists, but other units are used in some parts of the world. The metric system, which was developed in France in the late 18th century, is in everyday use in many countries, as well as being used by scientists; but imperial units (based on the traditional British measurement standard, also known as the foot–pound–second system), and standard units (based on commonly used American standards) are still in common use.

Whereas the basic units of length, mass and time were originally defined arbitrarily, scientists have sought to establish definitions of these which can be related to measurable physical constants; thus length is now defined in terms of the speed of light, and time in terms of the vibrations of a crystal of an particular atom. Mass, however, still eludes such definition, and is based on a piece of platinum-iridium metal kept in Sèvres, France.

☐ METRIC PREFIXES

Very large and very small units are often written using powers of ten; in addition the following prefixes are also used with SI units. Examples include: milligram (mg), meaning one thousandth of a gram, kilogram (kg), meaning one thousand grams.

Name	Number	Factor	Prefix	Symbol
trillionth	0.000000000001	10^{-12}	pico-	p
billionth	0.000000001	10^{-9}	nano-	n
millionth	0.000001	10^{-6}	micro-	μ
thousandth	0.001	10^{-3}	milli-	m
hundredth	0.01	10^{-2}	centi-	c
tenth	0.1	10^{-1}	deci-	d
one	1.0	10^{0}	–	–
ten	10	10^{1}	deca-	da
hundred	100	10^{2}	hecto-	h
thousand	1000	10^{3}	kilo-	k
million	1,000,000	10^{6}	mega-	M
billion	1,000,000,000	10^{9}	giga-	G
trillion	1,000,000,000,000	10^{12}	tera-	T
quadrillion	1,000,000,000,000,000	10^{15}	exa-	E

☐ CONVERSION FACTORS

Conversion of METRIC units to imperial (or standard) units

To convert:	to:	multiply by:
LENGTH		
millimeters	inches	0.03937
centimeters	inches	0.3937
meters	inches	39.37
meters	feet	3.2808
meters	yards	1.0936
kilometers	miles	0.6214
AREA		
square centimeters	square inches	0.1552
square meters	square feet	10.7636
square meters	square yards	1.196
square kilometers	square miles	0.3861
square kilometers	acres	247.1
hectares	acres	2.471
VOLUME		
cubic centimeters	cubic inches	0.061
cubic meters	cubic feet	35.315
cubic meters	cubic yards	1.308
cubic kilometers	cubic miles	0.2399
CAPACITY		
milliliters	fluid ounces	0.0351
milliliters	pints	0.00176 (0.002114 for US pints)
liters	pints	1.760 (2.114 for US pints)
liters	gallons	0.2193 (0.2643 for US gallons)
WEIGHT		
grams	ounces	0.0352
grams	pounds	0.0022
kilograms	pounds	2.2046
tonnes	tons	0.9842 (1.1023 for US, or short, tons)
TEMPERATURE		
Celsius	Fahrenheit	1.8, then add 32

Conversion of STANDARD (or imperial) units to metric units

To convert:	to:	multiply by:
LENGTH		
inches	millimeters	25.4
inches	centimeters	2.54
inches	meters	0.245
feet	meters	0.3048
yards	meters	0.9144
miles	kilometers	1.6094
AREA		
square inches	square centimeters	6.4516
square feet	square meters	0.0929
square yards	square meters	0.8316
square miles	square kilometers	2.5898
acres	hectares	0.4047
acres	square kilometers	0.00405
VOLUME		
cubic inches	cubic centimeters	16.3871
cubic feet	cubic meters	0.0283
cubic yards	cubic meters	0.7646
cubic miles	cubic kilometers	4.1678
CAPACITY		
fluid ounces	milliliters	28.5
pints	milliliters	568.0 (473.32 for US pints)
pints	liters	0.568 (0.4733 for US pints)
gallons	liters	4.55 (3.785 for US gallons)
WEIGHT		
ounces	grams	28.3495
pounds	grams	453.592
pounds	kilograms	0.4536
tons	tonnes	1.0161
TEMPERATURE		
Fahrenheit	Celsius	subtract 32, then $\times\,0.55556$

☐ SI UNITS

Now universally employed throughout the world of science and the legal standard in many countries, SI units (short for *Système International d'Unités*) were adopted by the General Conference on Weights and Measures in 1960. There are seven base units and two supplementary ones, which replaced those of the MKS (meter–kilogram–second) and CGS (centimeter–gram–second) systems that were used previously. There are also 18 derived units, and all SI units have an internationally agreed symbol.

None of the unit terms, even if named for a notable scientist, begins with a capital letter: thus, for example, the units of temperature and force are the kelvin and the newton (the abbreviations of some units are capitalized, however). Apart from the kilogram, which is an arbitrary standard based on a carefully preserved piece of metal, all the basic units are now defined in a manner that permits them to be measured conveniently in a laboratory.

Name	Symbol	Quantity	Standard
BASIC UNITS			
meter	m	length	The distance light travels in a vacuum in $\frac{1}{299,792,458}$ of a second
kilogram	kg	mass	The mass of the international prototype kilogram, a cylinder of platinum-iridium alloy, kept at Sèvres, France
second	s	time	The time taken for 9,192,631,770 resonance vibrations of an atom of cesium-133
kelvin	K	temperature	$\frac{1}{273.16}$ of the thermodynamic temperature of the triple point of water
ampere	A	electric current	The current that produces a force of 2×10^{-7} newtons per meter between two parallel conductors of infinite length and negligible cross section, placed one meter apart in a vacuum
mole	mol	amount of substance	The amount of a substance that contains as many atoms, molecules, ions or subatomic particles as 12 grams of carbon-12 has atoms
candela	cd	luminous intensity	The luminous intensity of a source that emits monochromatic light of a frequency 540×10^{-12} hertz and whose radiant intensity is $\frac{1}{683}$ watt per steradian in a given direction
SUPPLEMENTARY UNITS			
radian	rad	plane angle	The angle subtended at the center of a circle by an arc whose length is the radius of the circle
steradian	sr	solid angle	The solid angle subtended at the center of a sphere by a part of the surface whose area is equal to the square of the radius of the sphere

Name	Symbol	Quantity	Standard
DERIVED UNITS			
becquerel	Bq	radioactivity	The activity of a quantity of a radio-isotope in which 1 nucleus decays (on average) every second
coulomb	C	electric current	The quantity of electricity carried by a charge of 1 ampere flowing for 1 second
farad	F	electric capacitance	The capacitance that holds charge of 1 coulomb when it is charged by a potential difference of 1 volt
gray	Gy	absorbed dose	The dosage of ionizing radiation equal to 1 joule of energy per kilogram
henry	H	inductance	The mutual inductance in a closed circuit in which an electromotive force of 1 volt is produced by a current that varies at 1 ampere per second
hertz	Hz	frequency	The frequency of 1 cycle per second
joule	J	energy	The work done when a force of 1 newton moves its point of application 1 meter in its direction of application
lumen	lm	luminous flux	The amount of light emitted per unit solid angle by a source of 1 candela intensity
lux	lx	illuminance	The amount of light that illuminates 1 square meter with a flux of 1 lumen
newton	N	force	The force that gives a mass of 1 kilogram an acceleration of 1 meter per second per second
ohm	Ω	electric resistance	The resistance of a conductor across which a potential of 1 volt produces a current of 1 ampere
pascal	Pa	pressure	The pressure exerted when a force of 1 newton acts on an area of 1 square meter
siemens	S	electric conductance	The conductance of a material or circuit component that has a resistance of 1 ohm
sievert	Sv	dose	The radiation dosage equal to 1 joule equivalent of radiant energy per kilogram
tesla	T	magnetic flux density	The flux density (or density induction) of 1 weber of magnetic flux per square meter
volt	V	electric potential	The potential difference across a conductor in which a constant current of 1 ampere dissipates 1 watt of power
watt	W	power	The amount of power equal to a rate of energy transfer of (or rate of doing work at) 1 joule per second
weber	Wb	magnetic flux	The amount of magnetic flux that, decaying to zero in 1 second, induces an electromotive force of 1 volt in a circuit of one turn

The various quantities in physics are of necessity defined in terms of others, and the relationships between them may be used in experiments to determine actual values. For example, the density of an object is equal to its mass divided by its volume (because density equals mass per unit volume), whereas an object's weight is equal to its mass multiplied by the acceleration of free fall. The key relationships between physical quantities are shown here summarized by standard formulas.

PROPERTIES OF MATTER

Density d
$d = m/V$
    where m = mass, V = volume

Mass m
$m = F/a$
    where F = force, a = acceleration
$m = W/g$
    where W = weight,
    g = acceleration of free fall
$m = dV$
    where d = density, V = volume

Relative density RD
$RD = d_S/d_W$
    where d_S = density of substance,
    d_W = density of water

Weight W
$W = mg$
    where m = mass, g = acceleration
    of free fall

MECHANICS

Acceleration a
$a = F/m$
    where F = force, m = mass
$a = (v - u)/t$
    where v = final velocity,
    u = initial velocity, t = time

Displacement s
$s = vt$
    where v = velocity, t = time
$s = ut + \frac{1}{2}at^2$
    where u = initial velocity,
    a = acceleration, t = time
$s = (v^2 - u^2)/2a$
    where v = final velocity, u = initial
    velocity, a = acceleration

Distance ratio DR
    (= velocity ratio)
$DR = s_E/s_L$
    where s_E = speed of effort,
    s_L = speed of load

Efficiency E
$E = \dfrac{\text{work output}}{\text{work input}}$

$E = \dfrac{\text{force ratio (mechanical advantage)}}{\text{distance ratio (velocity ratio)}}$

Energy W
$W = Fs$
    where F = force, s = displacement

$W = mgh$
    where m = mass, g = acceleration
    of free fall, h = height
$W = \frac{1}{2}mv^2$
    where m = mass, v = velocity

Force F
$F = ma$
    where m = mass, a = acceleration
$F = P/v$
    where P = power, v = velocity
$F = W/s$
    where W = work (energy),
    s = displacement

Force ratio FR
    (= mechanical advantage)
FR = load/effort

Moment of a force
moment = Fs
    where F = force, s = displacement

Momentum
momentum = mv
    where m = mass, v = velocity
momentum = Ft
    where F = force, t = time

Power P
$P = W/t$
    where W = work (energy),
    t = time
$P = Fv$
    where F = force, v = velocity

Pressure p
$p = F/A$
    where F = force, A = area
$p = hdg$ for a liquid
    where h = height (depth),
    d = fluid density, g = acceleration
    of free fall
$p = kT/V$ for an ideal gas
    where k = Boltzmann constant,
    T = absolute temperature,
    V = volume

Speed S
$S = l/t$
    where l = distance, t = time

Velocity v
$v = s/t$
    where s = displacement, t = time
$v = u + at$
    where u = initial velocity,
    a = acceleration, t = time

$v^2 = u^2 + 2as$
    where u = initial velocity,
    a = acceleration, s = displacement

Work W
$W = Fs$
    where F = force, s = displacement

ELECTRICITY

Capacitance C
$C = Q/V$
    where Q = charge, V = voltage
$C = C_1 + C_2 + C_3 + ...$
    (for capacitors in parallel)
$1/C = 1/C_1 + 1/C_2 + 1/C_3 + ...$
    (for capacitors in series)

Charge Q
$Q = It$
    where I = current, t = time
$Q = CV$ for a capacitor
    where C = capacitance,
    V = voltage
$Q = F/E$ for a stationary charge
    where F = force, E = electric field
    strength

Current I
$I = V/R$
    where V = voltage, R = resistance
$I = Q/t$
    where Q = charge, t = time
$I = P/V$
    where P = power, V = voltage

Electric field strength E
$E = F/Q$
    where F = force, Q = charge
$E = V/s$
    where V = voltage, s = distance

Electromotive force $e.m.f.$
$e.m.f. = I/(R + r)$
    where I = current, $R + r$ = the total
    resistance of a circuit

Power P
$P = W/t$
    where W = electrical energy,
    t = time
$P = VI$
    where V = voltage, I = current
$P = I^2R$
    where I = current, R = resistance
$P = V^2/R$
    where V = voltage, R = resistance

Resistance R
$R = V/I$
    where V = voltage, I = current
$R = R_1 + R_2 + R_3 + ...$
    (for resistors in series)
$1/R = 1/R_1 + 1/R_2 + 1/R_3 + ...$
    (for resistors in parallel)

Voltage V
$V = IR$
    where I = current, R = resistance
$V = Q/C$ for a capacitor
    where Q = charge,
    C = capacitance
$V = W/Q$
    where W = (electrical) energy,
    Q = charge

HEAT

Specific latent heat L
$L = W/m$
    where W = energy, m = mass

Specific thermal capacity c
$c = W/m(T_2 - T_1)$
    where W = energy, m = mass,
    T_1 = initial temperature,
    T_2 = final temperature

Thermal capacity C
$C = W/(T_2 - T_1)$
    where W = energy, T_1 = initial
    temperature, T_2 = final
    temperature
$C = mc$
    where m = mass, c = specific
    thermal capacity

LIGHT

Focal distance f
$1/f = 1/v + 1/u$ for a lens
    where v = image distance,
    u = object distance
$f = 1/P$ for a lens
    where P = power
$f = r/2$ for a curved mirror
    where r = radius of curvature

Power (of a lens or mirror) P
$P = 1/f$
    where f = focal distance

Radius of curvature r
$r = 2f$ for a mirror
    where f = focal distance

Reflection
$i = r$
    where i = angle of incidence,
    r = angle of reflection

Refraction
$\sin i/\sin r = \mu$
    where i = angle of incidence,
    r = angle of refraction,
    μ = refractive index

Refractive index μ
$\mu = \sin i/\sin r$
    where i = angle of incidence,
    r = angle of refraction
$\mu = 1/\sin c$
    where c = critical angle

☐ PHYSICAL CONSTANTS

Certain constant quantities – such as the acceleration of free fall (the acceleration due to gravity) – recur in physical formulas and calculations. For this reason, they are known as physical (or fundamental, or universal) constants. They are assumed to be absolutely constant and therefore invariable throughout the Universe. The values here are given in SI units.

Many of the constants refer to properties of the three original elementary particles, the electron, neutron and proton, particularly their charges (if any) and their masses. Others describe the basic properties of matter, such as the Avogadro constant (the number of atoms or molecules in 1 mole of a substance) and the molecular volume of an ideal gas (the volume occupied by 1 mole of the gas). Loschmidt's constant is the number of particles in a cubic meter of an ideal gas at standard temperature and pressure (equal to one atmosphere, the pressure exerted by a 760mm column of mercury at 0°C).

Other constants are the proportionality constants in various fundamental equations. Examples include the gravitational constant G, which arises from an expression of Newton's law of gravitation: $F = G\,(m_1 m_2/d^2)$, where F is the gravitational force of attraction between masses m_1 and m_2 located at a distance d apart. Similarly the gas constant R appears in the universal gas equation: $pV = nRT$, where p, V and T are the pressure, volume and absolute temperature of an amount n of an ideal gas.

The Boltzmann constant is the gas constant divided by the Avogadro constant, and the Faraday constant is the electric charge carried by 1 mole of electrons or ions carrying a single charge, equal in turn to the Avogadro constant multiplied by the electron charge. The Rydberg constant occurs in formulas that describe the spectra of atoms, and is an indication of the binding energy of electrons.

Constant	Symbol	Value (SI units)
Acceleration of free fall	g	9.80665 m s^{-2}
Avogadro constant	N_A	6.02252×10^{26} kmol^{-1}
Bohr magnetron	μ_B	9.2732×10^{-27} J T^{-1}
Bohr radius	a_0	5.2917×10^{-11} m
Boltzmann constant	k	1.38054×10^{-23} J K^{-1}
Compton wavelength		
of electron	λ_c	2.4263×10^{-12} m
of neutron	λ_{cn}	1.3195×10^{-15} m
of proton	λ_{cp}	1.3214×10^{-15} m
Electric constant	ε_0	8.854×10^{-12} F m^{-1}
(permittivity of free space)		
Electron charge	e	1.60210×10^{-19} C
Electron charge/mass ratio	e/m_e	1.758796×10^{11} C kg^{-1}
Electron magnetic moment	μ_e	9.2848×10^{-24} J T^{-1}
Electron/proton magnetic		
moment ratio	μ_e/μ_p	6.58210×10^2
Electron radius	r_e	2.81777×10^{-15} m
Electron rest mass	m_e	9.10908×10^{-31} kg
Faraday constant	F	9.64870×10^4 C mol^{-1}
First radiation constant	c_1	3.7415×10^{-16} W m^2

Constant	Symbol	Value (SI units)
Gas constant	R_0	8.31434 J K^{-1} mol^{-1}
Gravitational constant	G	6.670×10^{-11} m^2 kg^{-2}
Loschmidt's constant	N_L	2.6872×10^{25} m^{-3}
Magnetic constant	μ_0	$4\pi \times 10^{-7}$ H m^{-1}
(permeability of free space)		
Mass unit	u	1.66043×10^{-27} kg
Molecular volume		
of an ideal gas at s.t.p.	V_m	22.414×10^{-3} m^3 mol^{-1}
Muon rest mass	m_μ	1.8836×10^{-28} kg
Neutron rest mass	m_n	1.67482×10^{-27} kg
Planck constant	h	6.62559×10^{-34} J s
Proton/electron mass ratio	m_p/m_r	1.8362×10^3
Proton magnetic moment	μ_p	1.41062×10^{-26} J T^{-1}
Proton rest mass	m_p	1.67252×10^{-27} kg
Rydberg constant	R	1.097373×10^7 m^{-1}
Second radiation constant	c_2	1.4388×10^{-2} m K
Speed of light	c	2.997925×10^8 m s^{-1}
Standard atmosphere		101325 N m^{-2}
Stefan-Boltzmann constant	σ	5.6697×10^{-8} W m^{-2} K^{-4}

☐ ELEMENTARY PARTICLES

Elementary (or fundamental) particles are the basic constituents of all matter. All are subatomic, in the sense that they are smaller than, or constituents of, atoms. The original elementary particles – the first to be discovered – were the electron (a lepton) and the proton and neutron (both baryons). Other leptons include the muon, tauon, three types of neutrino and their respective antiparticles.

According to current theory, the most fundamental of all are quarks, which make up hadrons (baryons and mesons). There are 12 types of quarks altogether: six quarks and six antiquarks. The six are distinguished by their *flavor*: up (u), down (d), charm (c), strange (s), top (t) [questionable] or bottom (b). In the following table, the symbols of antiquarks are distinguished by underlining: u, d, c, s, and b.

Type	Particle	Mass (MeV/c^2)	Isotopic spin	Quark content
Baryons	proton	938.3	½	uud
	neutron	939.6	½	udd
	lambda	1115.6	0	uds
	sigma-plus	1189.4	1	uus
	sigma-zero	1192.5	1	uds
	sigma-minus	1197.4	1	dds
	xi-zero	1314.9	½	uss
	xi-minus	1321.3	½	dss
	omega-minus	1672.5	0	sss
	lambda-plus	2285	0	udc
Gauge bosons	photon	0		
	W-plus/minus	80000		
	Z-zero	91000		

Type	Particle	Mass (MeV/c^2)	Isotopic spin	Quark content
Leptons	neutrino	0		
	electron	0.511		
	muon	105.6		
	tauon	1784		
Mesons	pi-plus/minus	139.6	1	ud, ud
	pi-zero	135.0	1	uu, dd
	K-plus/minus	493.7	½	us, su
	K-zero	497.7	½	ds, ds
	eta-zero	548.8	0	uu, dd, ss
	D-plus/minus	1869.4	½	cd, dc
	D-zero	1864.7	½	cu
	B-plus/minus	5270.8	½	ub, bu
	B-zero	5274.2	½	db

There are 92 naturally occurring elements; another 13 or so have been produced artifically. An element's identity is defined by its atomic number, which is the number of protons in its atomic nucleus (often given the abbreviation Z). The number of neutrons in the nucleus can vary; atoms with the same number of protons but different numbers of neutrons are isotopes of the same element.

Natural elements are also characterized by their relative atomic mass (or atomic weight), which is the average mass of the atoms (taking into account its isotopes) related to the mass of an atom of the isotope of carbon of mass 12 (the existence of heavier isotopes of carbon such as carbon-14 means that carbon's own relative atomic mass is more than 12). The masses given for radioactive elements are those of the longest-lived isotope.

Name of element	Chemical symbol	Atomic number	Relative atomic mass
Actinium	Ac	89	(227)
Aluminum	Al	13	26.9815
Americium	Am	95	(243)
Antimony	Sb	51	121.75
Argon	Ar	18	39.948
Arsenic	As	33	74.9216
Astatine	At	85	(210)
Barium	Ba	56	137.34
Berkelium	Bk	97	(247)
Beryllium	Be	4	9.0122
Bismuth	Bi	83	208.9806
Boron	B	5	10.81
Bromine	Br	35	79.904
Cadmium	Cd	48	112.40
Calcium	Ca	20	40.0
Californium	Cf	98	(251)
Carbon	C	6	12.001
Cerium	Ce	58	140.12
Cesium	Cs	55	132.9055
Chlorine	Cl	17	35.453
Chromium	Cr	24	51.996
Cobalt	Co	27	58.9332
Copper	Cu	29	63.546
Curium	Cm	96	(247)
Dysprosium	Dy	66	162.50
Einsteinium	Es	99	(254)
Erbium	Er	68	167.26
Europium	Eu	63	151.96
Fermium	Fm	100	(257)
Fluorine	F	9	18.9984
Francium	Fr	87	(223)
Gadolinium	Gd	64	157.25
Gallium	Ga	31	69.72
Germanium	Ge	32	72.59
Gold	Au	79	196.9665
Hafnium	Hf	72	178.49
Hahnium	Ha	105	–
Helium	He	2	4.0026
Holmium	Ho	67	164.9303
Hydrogen	H	1	1.0080
Indium	In	49	114.82
Iodine	I	53	126.904
Iridium	Ir	77	192.22
Iron	Fe	26	55.847
Krypton	Kr	36	83.80
Lanthanum	La	57	138.9055
Lawrencium	Lr	103	(257)
Lead	Pb	82	207.19
Lithium	Li	3	6.941
Lutetium	Lu	71	174.97
Magnesium	Mg	12	24.305
Manganese	Mn	25	54.9380
Mendelevium	Md	101	(258)
Mercury	Hg	80	200.59
Molybdenum	Mo	42	95.94
Neodymium	Nd	60	144.24
Neon	Ne	10	20.179
Neptunium	Np	93	(237)
Nickel	Ni	28	58.71
Niobium	Nb	41	92.9064
Nitrogen	N	7	14.0067
Nobelium	No	102	(255)
Osmium	Os	76	190.2
Oxygen	O	8	15.9994
Palladium	Pd	46	106.4
Phosphorus	P	15	30.9738
Platinum	Pt	78	195.09
Plutonium	Pu	94	(244)
Polonium	Po	84	(209)
Potassium	K	19	39.102
Praseodymium	Pr	59	140.9077
Promethium	Pm	61	(145)
Protactinium	Pa	91	231.0359
Radium	Ra	88	226.0254
Radon	Rn	86	(222)
Rhenium	Re	75	186.20
Rhodium	Rh	45	102.9055
Rubidium	Rb	37	85.4678

METALS AND NONMETALS

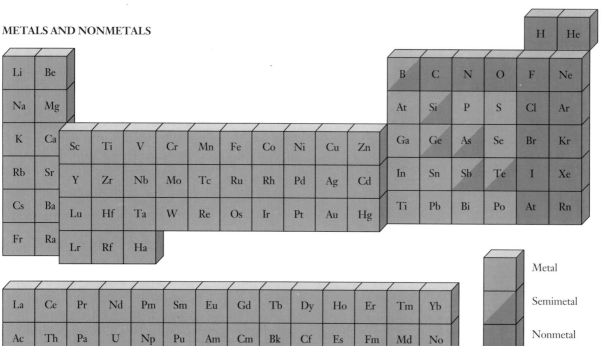

◄ Most elements are metals – generally hard substances that are good conductors of heat and electricity. Many of the nonmetals are gases, which can conduct electricity when they are in the form of ions at low pressures. Six elements – including silicon and germanium – are termed semimetals or metalloids. Most of these are semiconductors.

PHASES OF THE ELEMENTS

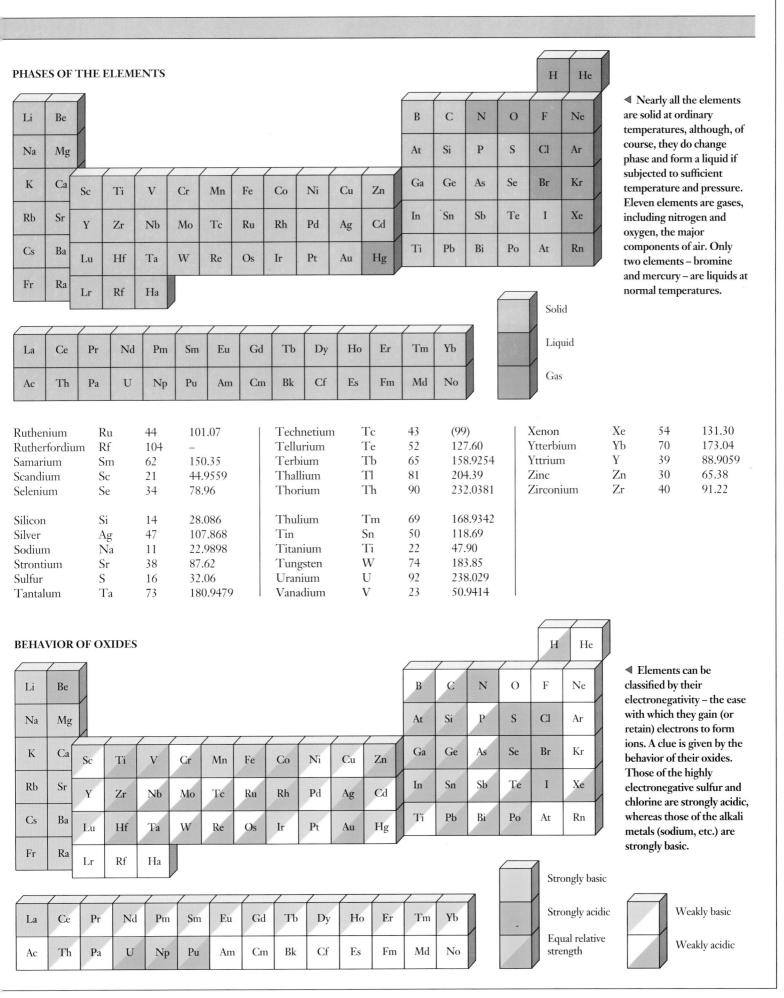

◄ Nearly all the elements are solid at ordinary temperatures, although, of course, they do change phase and form a liquid if subjected to sufficient temperature and pressure. Eleven elements are gases, including nitrogen and oxygen, the major components of air. Only two elements – bromine and mercury – are liquids at normal temperatures.

Solid
Liquid
Gas

Ruthenium	Ru	44	101.07		Technetium	Tc	43	(99)
Rutherfordium	Rf	104	–		Tellurium	Te	52	127.60
Samarium	Sm	62	150.35		Terbium	Tb	65	158.9254
Scandium	Sc	21	44.9559		Thallium	Tl	81	204.39
Selenium	Se	34	78.96		Thorium	Th	90	232.0381
Silicon	Si	14	28.086		Thulium	Tm	69	168.9342
Silver	Ag	47	107.868		Tin	Sn	50	118.69
Sodium	Na	11	22.9898		Titanium	Ti	22	47.90
Strontium	Sr	38	87.62		Tungsten	W	74	183.85
Sulfur	S	16	32.06		Uranium	U	92	238.029
Tantalum	Ta	73	180.9479		Vanadium	V	23	50.9414

Xenon	Xe	54	131.30
Ytterbium	Yb	70	173.04
Yttrium	Y	39	88.9059
Zinc	Zn	30	65.38
Zirconium	Zr	40	91.22

BEHAVIOR OF OXIDES

◄ Elements can be classified by their electronegativity – the ease with which they gain (or retain) electrons to form ions. A clue is given by the behavior of their oxides. Those of the highly electronegative sulfur and chlorine are strongly acidic, whereas those of the alkali metals (sodium, etc.) are strongly basic.

Strongly basic
Strongly acidic
Equal relative strength

Weakly basic
Weakly acidic

Metals are among the most versatile materials in use. The reason is their extremely wide range of physical properties. For example, relative densities range from 22 for osmium and iridium to less than 1 for sodium and potassium. Cesium and gallium melt at temperatures below that of the human body, whereas tungsten melts at a temperature in excess of 5500°C. In contrast, the strange metal mercury does not become solid unless it is cooled below -38.8°C. Hardness, resistivity and thermal capacity also vary widely, giving engineers a broad range of materials to suit a particular application, although figures for some of the properties are not readily available (as shown by the gaps on the table below).

	Relative density	Melting point (°C)	Boiling point (°C)	Resistivity at 0°C (10^{-8} ohm m)	Thermal conductivity at 0°C (W cm^{-1} K^{-1})	Hardness (Mohs scale)
Aluminum	2.70	660.5	2520	2.42	2.36	2–2.9
Barium	3.51	729	1805	30.2		
Beryllium	1.85	1289	2472	3.02		
Bismuth	9.78	271.4	1564			2.5
Cadmium	8.65	321	767			2.0
Calcium	1.54	842	1494	3.11		1.5
Cesium	1.88	28.4	671	18.7		0.2
Chromium	7.20	1863	2672	11.8		9.0
Cobalt	8.90	1495	2928	6.35		
Copper	1.54	1085	2563	1.54	4.01	2.5–3
Gallium	5.90	29.8	2205			1.5
Germanium	5.47	938	2834			
Gold	19.31	1064	2857	2.05	3.18	2.5–3
Hafnium	13.31	2231	4603	30.4		
Iridium	22.4	2447	4428	6.0		6–6.5
Iron	7.86	1538	2862	8.57	0.84	4–5
Lead	11.34	327.5	1750	19.2		1.5
Lithium	0.53	181	1342	4.05		0.6
Magnesium	1.74	650	1090	4.05		2.0
Manganese	7.20	1246	2062	5.0		5.0
Mercury	13.6	−38.8	357	95.8		
Molybdenum	10.20	2623	4639	4.85		
Nickel	8.90	1455	2914	6.16		
Osmium	22.48	3033	5012			7.0
Palladium	12.02	1555	2964	9.78		4.8
Platinum	21.45	1769	3827	9.6	0.73	4.3
Plutonium	19.84	640	3230			
Potassium	0.86	63.7	759	6.49		0.5
Radium	5.0	700	1140			
Rhodium	12.1	1963	3697			
Rubidium	1.48	39.5	688	11.5		0.3
Selenium	4.45	221	685			2.0
Silver	10.50	962	2163	1.47	4.28	2.5–4
Sodium	0.97	97.8	883	4.33		0.4
Strontium	2.62	769	1382	12.3		1.8
Tantalum	16.62	3020	5458	12.1		
Tin	7.3	232	2603			1.5–1.8
Titanium	4.5	1670	3289			
Tungsten	19.35	3422	5555	4.82	1.82	
Uranium	18.68	1135	4134			
Vanadium	5.96	1910	3409	18.1		
Zinc	7.14	419.6	907	5.46		2.5
Zirconium	6.49	1855	4409	38.8		

Physicists, engineers and draftsmen use various standard symbols to represent electrical or electronic devices in circuit diagrams. Some of the most commonly employed symbols are shown here.

Device	Symbol	Device	Symbol
Alternating current		NOT gate (inverter)	
Alternating current pulse		npn transistor	
Ammeter		OR gate	
Amplifier		Piezoelectric crystal	
AND gate		Photodiode	
Antenna		Plug and socket	
Battery or accumulator		pnp transistor	
Capacitor		Positive-going pulse	
Connection of conductors		Positive-going step function	
Earth (ground)		Primary cell (longer line is +)	
Electromagnetic effect		Resistor	
Exclusive OR gate		Resistor with sliding contact	
Fault		Semiconductor diode	
Inductor (coil or choke)		Switch	
Inductor with magnetic core		Terminal	
Junction of conductors		Thermal effect	
Light-emitting diode		Transformer	
Negative-going pulse		Variability (noninherent)	
Negative-going step function		Variability (stepwise)	
NAND gate (negated AND)		Variable resistor	
NOR gate (negated OR)		Voltmeter	

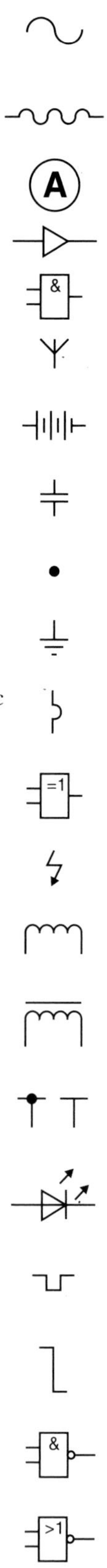

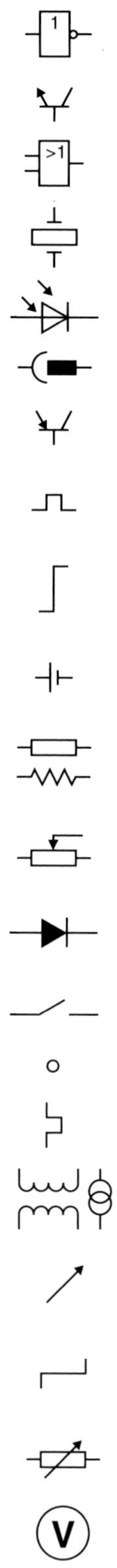

□ SOUND INTENSITY

The pitch of a sound is determined by the frequency of its waves, measured in hertz (Hz), and the loudness is primarily a function of their intensity. The unit of intensity of sound is the decibel (dB). The zero point of the decibel scale is defined as the "hearing threshold", the quietest sound that a normal human ear can detect; and other sounds are measured relative to this level. The scale is logarithmic, so that a 10dB sound is ten times the intensity, a 20dB sound is 100 the intensity, and a 30dB sound is 1000 times the intensity of a 0dB sound. A sound is normally considered painful at about 90dB, and physically damaging at 100dB.

The sounds that a typical human ear can perceive depend not just on their loudness, but also on their pitch. Most people are most sensitive to sounds of around 5000Hz. However, the hearing abilities of the ear decline and the higher frequency sounds, irrespective of their loudness, become increasingly less audible with age.

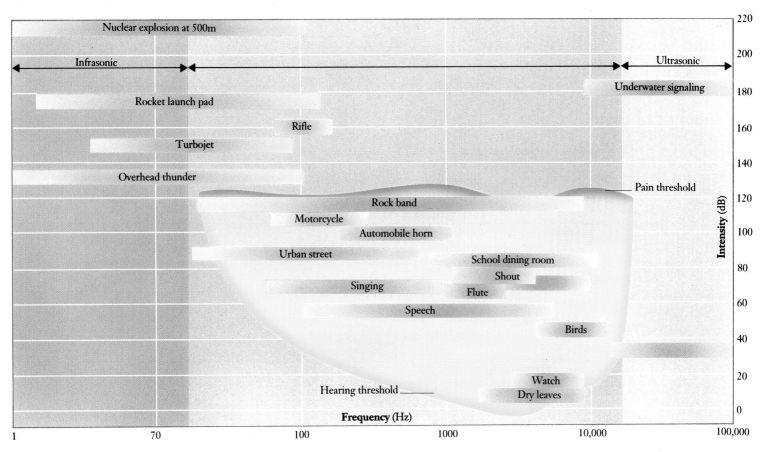

□ FURTHER READING

Atkins, P.W. *Molecules* (New York, 1983)

Barger, V. and Phillips, R.J. *Collider Physics* (Reading, 1987)

Beiser, A. *Physics* (Menlo Park, 4th ed 1986)

Berger, C.R. and Chafee, S.H. (editors) *Handbook of Communications Science* (San Mateo, 1987)

Calder, N. *Einstein's Universe* (London, 1976)

Chant, C. (editor) *Science, Technology and Everyday Life* (London, 1989)

Claydon, W.R. *Matter and Spirit* (New York, 1981)

Cline, B.L. *Men who Made a New Physics* (Chicago, 1987)

Dacre, J. *Nuclear Physics* (London, 1980)

Fielding, R. (editor) *A Technological History of Motion Pictures and Television* (Berkeley, 1967)

Gibbs, K. *Advanced Physics* (Cambridge, 2nd ed. 1990)

Goodstein, D.L. *States of Matter* (Englewood Cliffs, 1975)

Hellemans, A. and Bunch, B. *The Timetables of Science* (New York, 1988)

Hey, T. and Walters, P. *The Quantum Universe* (Cambridge, 1987)

Kinsler, L.E. et al. *Fundamentals of Acoustics* (New York, 1982)

Krakovitz, R. *High Energy* (Los Angeles, 1986)

Longstaff, M. *Unlocking the Atom: A Hundred Years of Nuclear Energy* (London, 1980)

Marshall, S.V. and Skitek, G.G. *Electromagnetic Concepts and Applications* (Englewood Cliffs, 1987)

Marton, L. *Early History of the Electron Microscope* (San Francisco, 1968)

Mock, D.E. and Vargish, T. *Inside Relativity* (Princeton, 1987)

Morris, R. *The Nature of Reality* (New York, 1987)

Olenik et al. *The Mechanical Universe: Introduction to Mechanics and Heat* (New York, 1985)

Page, R.M. *The Origin of Radar* (New York, 1962)

Polkinghorne, W.G. and Scott, P.R. *Structure and Spectra of Molecules* (New York, 1985)

Sabra, A.I. *Theories of Light from Descartes to Newton* (New York, 1981)

Segré, E. *From X Rays to Quarks* (New York, 1980)

Seymour, J. *Electronic Devices and Components* (New York, 1981)

Sobel, M.I. *Light* (Chicago, 1987)

Walker, P. (editor) *Chambers Science and Technology Dictionary* (Edinburgh, 1988)

Weinberg, S. *The Discovery of Subatomic Particles* (New York, 1983)

Wober, Mallory and Gunter *Television and Social Control* (New York, 1988)

Most of the properties of the elements – and hence their behavior in physics – depend on their stability. This, in turn, is a function of their atomic make-up, particularly of the numbers of protons and neutrons in their nuclei. The chart below plots the numbers of protons (equal to the atomic number, Z) against the number of neutrons for about 300 atoms that occur on Earth.

Many are isotopes – forms of an element that have the same atomic number but differ in the number of neutrons in their nuclei. With the exception of hydrogen, the stable isotopes of the light elements, up to calcium ($Z = 20$), fall on a line representing equal Z and N. Thereafter the stability line curves away as nuclei contain more neutrons than protons (they have a higher N than Z).

The chart shows that the naturally occurring isotopes form an almost complete sequence from hydrogen ($Z = 1$) to uranium ($Z = 92$). But there are two gaps, corresponding to technetium ($Z = 43$) and promethium ($Z = 61$), which are synthetic elements. They are made in particle accelerators or turn up in the fission products of nuclear explosions. They each have 15 or more isotopes, all of which are radioactive. They have such short half-lives that any that once existed on Earth have long ago decayed to form other elements.

Some elements, usually those with odd atomic numbers, have only one isotope; examples include fluorine ($Z = 9$), manganese ($Z = 25$), arsenic ($Z = 33$), iodine ($Z = 53$) and bismuth ($Z = 83$), the heaviest nonradioactive element. Their stability seems to depend on whether they contain an even or odd number of nuclear particles. In fact, there are many more stable isotopes with even numbers of protons and neutrons. Very few have odd numbers of protons *and* neutrons, including deuterium (an isotope of hydrogen with atomic number 1 and neutron number 1), lithium ($Z = 3$, $N = 3$), boron ($Z = 5$, $N = 5$) and nitrogen ($Z = 7$, $N = 7$).

Not all naturally occurring isotopes are stable. Some have combinations of Z and N that cannot permanently remain together in a nucleus. Their nuclei achieve stability by emitting a particle or particles – they are radioactive. In one such mechanism, the nucleus loses a combination of two protons and two neutrons, known as an alpha particle or a helium nucleus. This has the effect of lowering both Z and N by 2. In another form of radioactive decay, the unstable nucleus converts one of its neutrons into a proton and an electron (which is emitted as a beta particle). This has the effect of increasing Z by 1. A few nuclei emit both alpha and beta particles to form two new kinds of atoms at the same time. The naturally occurring radioactive isotopes lighter than bismuth are named on the chart. In addition to technetium and promethium, all elements beyond bismuth are radioactive and belong to one of the three natural decay series (*see* opposite page).

Naturally-occurring isotopes

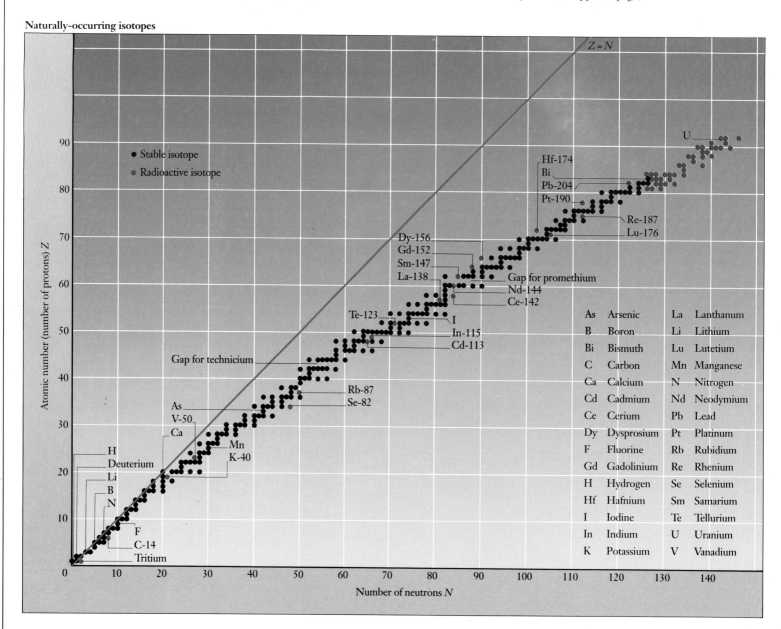

Thorium-232, actinium-235 and uranium-238 are three heavy radioisotopes which occur naturally on Earth. They continue to exist because each has an extremely long half-life – the time it takes for half of its atoms to decay (by emitting radioactivity) to form other atoms. For these three isotopes, the half lives are 14 billion, 7 billion and 4.5 billion years, respectively. They decay by successively emitting alpha particles (helium nuclei, mass 4) or beta particles (electrons, negligible mass) until they eventually form stable (nonradioactive) isotopes of lead. These three decay series are known as the natural decay series. There is also a fourth, beginning with the artifically produced plutonium-241, which does not occur naturally, and ends with a stable isotope of bismuth.

Uranium series

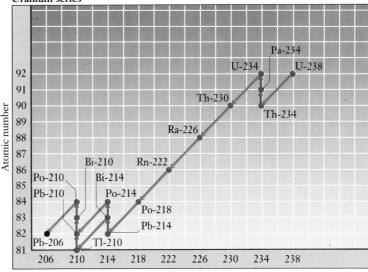

Thorium series

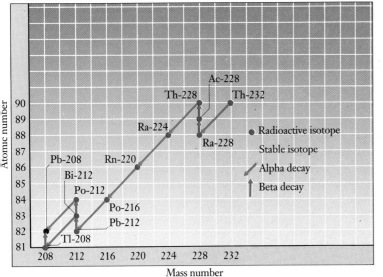

 The thorium decay series involves 12 stages. It is known as the 4*n* series, because each isotope has a mass number divisible by 4. It starts with thorium-232 and ends with lead-208. One of the intermediates is a gas, radon-220. Its half-life is 54 seconds.

95 Americium (Am)	87 Francium (Fr)
94 Plutonium (Pu)	86 Radon (Rn)
93 Neptunium (Np)	85 Astatine (At)
92 Uranium (U)	84 Polonium (Po)
91 Protactinium (Pa)	83 Bismuth (Bi)
90 Thorium (Th)	82 Lead (Pb)
89 Actinium (Ac)	81 Thallium (Tl)
88 Radium (Ra)	

▲ The uranium decay series, starting with uranium-238, ends with lead-206. It is called the 4*n* + 2 series, because each individual isotopic mass in the series gives a remainder of 2 when divided by 4. Some of the intermediate isotopes are extremely shortlived (for example, polonium-214 has a half-life of less than one thousandth of a second).

Plutonium series

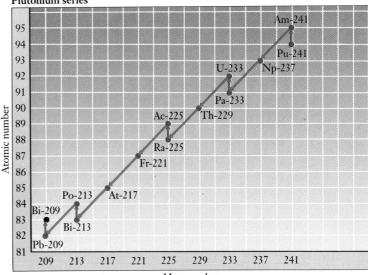

Actinium series

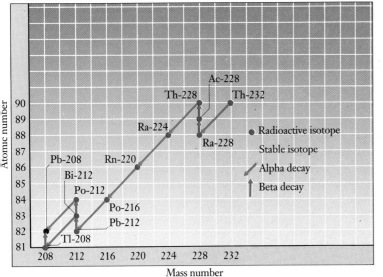

◄ The actinium decay series, named after actinium-277 (which was the first member of the series to be discovered), actually starts with uranium-235. Through many changes, some of them branched, it ends at the stable isotope lead-207. It is also known as the 4*n* + 3 series because each member isotope has a mass that is a multiple of 4 + 3.

▲ The "artificial" plutonium series (which was originally known as the neptunium series after its longest-lived member) is also called the 4*n* + 1 series because the atomic masses of each of its members yield a remainder of 1 when divided by 4. All of the lighter isotopes – those with masses less than 229 – have half-lives of less than 15 days.

The Nobel Prize for Physics, like similar prizes for Chemistry, Physiology or Medicine, Literature and Peace, has been awarded annually since 1901 as the world's most prestigious prize for outstanding work in the field, under the terms of the will of the Swedish chemist and engineer Alfred Nobel, who died in 1896. The prizes for physics and for chemistry are awarded by the Royal Swedish Academy of Sciences.

1901 **Wilhelm Röntgen** *German*
Discovery of X rays

1902 **Hendrik Lorentz and Pieter Zeeman** *Dutch*
Investigating the effect of magnetic fields on light (Zeeman effect)

1903 **Pierre and Marie Curie and Henri Becquerel** *French*
Discovery of radioactivity

1904 **Lord Rayleigh** *British*
Discovery of argon

1905 **Phillip Lenard** *German*
Study of cathode rays

1906 **Joseph John Thomson** *British*
Investigation of gaseous discharges

1907 **Albert Michelson** *American*
Measuring the speed of light

1908 **Gabriel Lippmann** *French*
Development of color photography (in spectrography)

1909 **Guglielmo Marconi** *Italian* **and Ferdinand Braun** *German*
Development of wireless telegraphy

1910 **Johannes van der Waals** *Dutch*
Study of intermolecular forces in liquids and gases

▲ **Wilhelm Röntgen, the first ever winner of the Nobel prize in 1901.**

1911 **Wilhelm Wien** *German*
Investigation of black body radiation

1912 **Nils Dalen** *Swedish*
Invention of automatic gas supply for lighthouses

1913 **Heike Kamerlingh-Onnes** *Dutch*
Investigation of the liquefaction of gases

1914 **Max von Laue** *German*
Study of X rays by crystal diffraction

1915 **William Henry Bragg and William Lawrence Bragg** *British*
Study of crystal structure using X rays

1916 No award

1917 **Charles Barkla** *British*
Study of X-ray and other short-wavelength emission

1918 **Max Planck** *German*
Proposal of the quantum theory of the atom

1919 **Johannes Stark** *German*
Study of spectra in electric fields (the Stark effect)

1920 **C. Gauillaume** *Swiss*
Development of low-expansion nickel alloys

1921 **Albert Einstein** *German/Swiss*
Explanation of the photoelectric effect and contributions to mathematical physics

1922 **Niels Bohr** *Danish*
Theories about atomic structure and atomic spectra

1923 **Robert Millikan** *American*
Description of the photoelectric effect and determination of the electronic charge

1924 **Karl Siegbahn** *Swedish*
Development of the technique of X-ray spectroscopy

1925 **James Franck and G. Hertz** *German*
Explanation of the interaction between electrons and atoms (in quantum theory)

1926 **Jean-Baptiste Perrin** *French*
Research into the sizes of atoms and molecules and thus the structure of matter

1927 **Arthur Compton** *American* **and Charles (C.T.R.) Wilson** *British*
Explanation of X-ray reflection (Compton effect) and the development of the cloud chamber

1928 **Owen Richardson** *British*
Explanation of thermionic emission

1929 **Louis de Broglie** *French*
Work on the wave nature of matter

1930 **Chandrasekhara Raman** *Indian*
Researches into the scattering of radiation by media (Raman effect)

1931 No award

▲ **Max Planck, German pioneer of quantum physics, prizewinner in 1918**

1932 **Werner Heisenberg** *German*
Development (with Max Born) of quantum mechanics, and the proposal of the uncertainty (Heisenberg) principle

1933 **Paul Dirac** *English* **and Erwin Schrödinger** *Austrian*
Devolopment of wave mechanics (the mathematics of quantum theory)

1934 No award

1935 **James Chadwick** *British*
Discovery of the neutron

1936 **Carl Anderson** *American* **and Victor Hess** *Austrian-born American*
Discovery of the positron and investigation of cosmic rays

1937 **Clinton Davisson** *American* **and George Thomson** *British*
Discovery of electron diffraction by crystals

1938 **Enrico Fermi** *Italian*
Production of transuranium elements

1939 **Ernest Lawrence** *American*
Invention of the cyclotron

1940-42 No award

1943 **Otto Stern** *German-born American*
Development of molecular beams (for studying subatomic particles)

1944 **Isidor Rabi** *American*
Study of magnetic properties of atomic nuclei

1945 **Wolfgang Pauli** *Austrian-Born Swiss*
Formulation of the exclusion principle

1946 **Percy Bridgman** *American*
Work in high-pressure physics and thermodynamic

1947 **Edward Appleton** *British*
Study of the ionosphere

1948 **Patrick (P.M.S.) Blackett** *British*
Study of cosmic rays

1949   **Hideki Yukawa** *Japanese*
Prediction of the existence of the meson

1950   **Cecil Powell** *British*
Photographic study of atomic nuclei and the discovery of the meson

1951   **John Cockcroft** *British* **and Ernest Walton** *Irish*
Use of accelerated particles (protons) to transmute nuclei

1952   **Franz Bloch** *Swiss/German-born American* **and Edward Purcell** *American*
Study of atomic energy levels and nuclear magnetic resonance (NMR)

1953   **Fritz Zernike** *Dutch*
Development of phase-contrast microscopy

1954   **Max Born and Walther Bothe** *German*
Contributions to quantum mechanics and the study of electron emission

1955   **Willis Lamb and Polycarp Kusch** *American*
Study of hydrogen spectra and the precise determination of the magnetic moment of the electron

1956   **John Bardeen, William Brattan and William Shockley** *American*
Development of the transistor

1957   **Tsung Dao Lee and Chen Ning Yang** *Chinese-born American*
Disproof of the conservation of parity law

1958   **Pavel Cherenkov, I. Frank and I. Tamm** *Soviet*
Studies of the beahvior of high-energy particles and the Cherenkov effect

1959   **Emilio Segrè** *Italian-born American* **and Owen Chamberlain** *American*
Proving the existence of the antiproton

1960   **Donald Glaser** *American*
Invention of the bubble chamber

1961   **Rolf Hofstadter** *American* **and Rudolf Mössbauer** *German-born American*
Study of nucleons and gamma rays with repsect to atomic structure

1962   **Lev Landau** *Soviet*
Researches into liquid helium (and its implications to the states of matter at low temperatures)

1963   **Eugene Wigner** *Hungarian-born American*, **Maria Goeppert-Mayer** *German-born American* **and Hans Jensen** *German*
Original studies of into the structures of atomic nuclei

1964   **Charles Townes** *American*, **Nikolai Basov and Alexander Prokhorov** *Soviet*
Development of masers and lasers

1965   **Richard Feynman, Julian Schwinger** *American* **and Sinichiro Tomonaga** *Japanese*
Studies in quantum electrodynamics

1966   **Alfred Kastler** *French*
Study of atomic energy levels and their role in lasers

1967   **Hans Bethe** *American*
Proposition of the theory of nuclear reactions

1968   **Luis Alvarez** *American*
Study of subatomic particles

1969   **Murray Gell-Mann** *American*
Classification of nuclear particles and the idea of quarks

1970   **H. Alfvén** *Swedish* **and L. Néel** *French*
Work on magnetohydrodynamics and magnetic computer memories

1971   **Dennis Gabor** *Hungarian-born British*
Development of holography

1972   **John Bardeen, Leon Cooper and John Schrieffer** *American*
Study of superconductivity

1973   **Ivar Giaever** *American*, **Leo Esaki** *Japanese* **and B. Jesephson** *British*
Discovery of the tunneling effect in semiconductors

1974   **A. Hewish and Martin Ryle** *British*
Discovery of pulsars and advances in radioastronomy

1975   **J. Rainwater** *American*, **A. Bohr and B. Mottelson** *Danish*
Theory of atomic nuclear structure

1976   **B. Richter and S. Ting** *American*
Discovery of the psi particle

1977   **Philip Anderson, John Van Vleck** *American* **and Neville Mott** *British*
Semiconductor development

▲ **Werner Heisenberg, quantum physicist and prizewinner in 1932.**

1978   **Pyotr Kapitsa** *Soviet*, **Arno Penzias and Robert Wilson** *American*
Low-temperature studies and discovery of cosmic background radiation

1979   **Sheldon Glashow, Steven Weinberg** *American* **and Abdus Salam** *Pakistani*
Development of unified field theory

1980   **James Cronin and Val Fitch** *American*
Disproof of laws of symmetry in subatomic particles

1981   **N. Bloembergen, A Schawlow** *American* **and K. Siegbahn** *Swedish*
Development of laser spectroscopy and high-resolution electron microscopy

1982   **Kenneth Wilson** *American*
Studies of changes of state

1983   **Subrahmanyan Chandrasekhar and William Fowler** *American*
Theories about the evolution and death of stars

1984   **Carlo Rubbia** *Italian* **and Simon van der Meer** *Dutch*
Discoveries of W particle and Z particle

1985   **Klaus von Klitzing** *German*
Electrical resistance measurement

1986   **Ernst Ruska, Gerd Binning** *German* **and Heinrich Rohrer** *Swiss*
Advances in the electron microscope and scanning tunelling microscope

1987   **G Bednorz** *German* **and K. Müller** *Swiss*
Studies of superconductivity in ceramics

1988   **Leon Lederman, Melvin Schwartz and Jack Steinberger** *American*
Discovery of neutrinos and development of their use in research

1989   **N. Ramsey, H. Dehmelt** *American* **and W. Paul** *German*
Development of cesium atomic clock

1990   **J. Friedman, H. Kendall** *American* **and R. Taylor** *Canadian*
Discovery of structures of neutrons and protons

1991   **P.-G. de Gennes** *French*
Analyses of alignments and orderly arrangements of molecules in certain substances

1992   **G. Charpak** *French*
Invention of particle detector

1993   **Russel A. Hulse and Joseph P. Taylor** *American*
Studies of binary pulsars and gravitational waves

INDEX

Figures in *italic type* refer to the captions to the illustrations, and to Timechart entries; figures in **bold type** refer to the Keywords and Factfile articles.

ACKNOWLEDGMENTS

Picture credits

1 Z 2–3t SPL/British Technical Films 2–3b AOL
4 SPL/Dr Mitsuo Ohtsuki 6 SPL/ Lockheed/Jisas
7t AOL 7b AOL 48-49 RF/Sipa Press/Kojo Tanaka
49 AOL 50-51 Z/ Kalt 54-55 UNEP-Select/Compoint
Stephane 56-57 AOL 57 Z/Horowitz 58-59 Z 62 SPL/
NASA 62-63 NASA 64-65 Spectrum Colour Library
65 HL/Sarah Errington 66 AOL 66-67 Z/Weir 68-69
Z 70-71 FSP/Liaison/ Kermani 71 FSP/Liaison/
Kermani 72bl AOL 72br AOL 72-73 Pilkington PLC
74-75 RF/ Rangefinders 75 AOL 77 SPL/Bruce Iverson
78-79 Nissan Sunderland 80-81 Z 81 Ron Boardman
83 IP/Homer Sykes 86 IP/Cosmos/G Buthand 88 AOL
90-91 AOL 92-93 AOL 94-95t RF/Geiss 94-95c SPL/
Manfred Kage 94-95b British Telecommunications
PLC 95t British Telecommunications PLC 95b AOL
96-97 TSW/Terry Vine 98-99 AOL 100-101 PEP/
James Hudnall 101 Topham Picture Source 102-103
NHPA/Stephen Dalton 103 SPL/CNRI 105 AOL 106-
107 Paul Brierley 107 AOL 108 AOL 108-109 Z/St
Westmorland 109 AOL 110 AOL 110-111 UNEP-
Select/Lau Chak Tao 111 Colorsport 112-113 AOL
113 UNEP-Select/Franz Franzler 114-115 Z/Starfoto
115 SPL/Dr Jeremy Burgess 117t RF/Sipa Press/Fraser
117b National Medical Slide Bank 118l SPL/Agfa-
Gevaert 118r SPL/British Technical Films 118-119
SPL/Dr R Clark & M Goff 119 SPL/Dr R J Allen
120 SPL/NASA 120-121 Spectrum Colour Library
123l SPL/Susan Leavines 123r ESA l993/ Radarsat
International 124-125 FSP/Paul Nightingale 125 Piran
Murphy 128-129 National Medical Slide Bank
129 AOL 132-133 Silkeborg Museum 133 SPL/
Philippe Plailly 134-135 SL/Dr Mitsuo Ohtsuki
135c Novosti London l992 135b Z/Bramaz 136
SPL/Roger Rossmeyer Starlight 136-137 SPL/
Lockheed/Jisas 137 SPL/US Department of Energy
138-139 Z/Weir 139 SPL/Alexander Tsiaras
141l SPL/Dr Jeremy Burges 141r SPL/David Scharf
142-143 SPL/Los Alamos National Laboratory
143 SPL/NASA/ Space Telescope Science Institute
154t Bundey Library, courtesy of AIP Niels Bohr
Library 154b Wellcome Institute Library, London
155 AOL

Abbreviations

b = bottom, **t** = top
l = left, **c** = center, **r** = right

AOL   Andromeda Oxford Limited, Abingdon, UK
FSP   Frank Spooner Pictures, London, UK
HL   Hutchison Library, London, UK
IP   Impact Photos, London, UK
PEP   Planet Earth Pictures, London, UK
RF   Rex Features, London, UK
SPL   Science Photo Library, London, UK
TSW   Tony Stone Worldwide, London, UK
Z   Zefa Picture Library, London, UK

Artists

Mike Badrock, John Davies, Hugh Dixon, Bill Donohoe,
Sandra Doyle, John Francis, Shami Ghale, Mick Gillah,
Ron Hayward, Jim Hayward, Trevor Hill/Vennor Art,
Joshua Associates, Frank Kennard, Pavel Kostell, Ruth
Lindsey, Mike Lister, Jim Robins, Colin Rose, Colin
Salmon, Leslie D. Smith, Ed Stewart, Tony Townsend,
Halli Verinder, Peter Visscher

Studio photography

Richard Clark

Editorial assistance

Richard Corfield, Peter Lafferty, Ray Loughlin

Index

Ann Barrett

Origination by

J Film, Bangkok; ASA Litho, UK